Exercises in Quantum Mechanics

Reidel Texts in the Mathematical Sciences

A Graduate-Level Book Series

Exercises in Quantum Mechanics

A Collection of Illustrative Problems and Their Solutions

Harry A. Mavromatis
Department of Physics, American University, Beirut, Lebanon

Springer-Science+Business Media, B.V.

Library of Congress Cataloging in Publication Data

Mavromatis, Harry A., 1941-
Exercises in quantum mechanics.

(Reidel Texts in the mathematical sciences)
Includes index.
1. Quantum theory–Problems, exercises, etc.
I. Title. II. Series.
QC174.15.M38 1986 530.1'2'076 86-25984

ISBN 978-94-015-7773-1 ISBN 978-94-015-7771-7 (eBook)
DOI 10.1007/978-94-015-7771-7

Originally published by D. Reidel Publishing Company, Dordrecht, Holland in 1987.
Softcover reprint of the hardcover 1st edition 1987

To Vasso:
οτι ηγαπησε πολυ

Table of Contents

Preface

This monograph is written within the framework of the quantum mechanical paradigm. It is modest in scope in that it is restricted to some observations and solved illustrative problems not readily available in any of the many standard (and several excellent) texts or books with solved problems that have been written on this subject. Additionally a few more or less standard problems are included for continuity and purposes of comparison.

The hope is that the points made and problems solved will give the student some additional insights and a better grasp of this fascinating but mathematically somewhat involved branch of physics.

The hundred and fourteen problems discussed have intentionally been chosen to involve a minimum of technical complexity while still illustrating the consequences of the quantum-mechanical formalism.

Concerning notation, useful expressions are displayed in rectangular boxes while calculational details which one may wish to skip are included in square brackets.

Beirut
June, 1985

HARRY A. MAVROMATIS

Schematic illustration of various approaches to calculating <u>Energy Levels</u> of

<u>Quantum Mechanical Systems</u>

<u>Generally useful approaches:</u>

1) Schrödinger Equation in Momentum Space (Chapter 3)

2) Schrödinger Equation in Coordinate Space (Chapters 8, 9, 12)

3) Poles of Scattering Amplitude (Chapter 6)

4) Perturbation Theory (Chapter 11)

5) Dalgarno-Lewis Technique (Chapter 14)

<u>High lying States:</u>

1) Wilson-Sommerfeld Quantization Condition (Chapter 1)

<u>Ground State:</u>

1) Uncertainty Principle (Chapter 5)

2) Variational Approach (Chapter 10)

CHAPTER 1

Wilson-Sommerfeld Quantization Condition

The hydrogen atom, when treated using Bohr's admixture of classical and quantum concepts involves an electron circulating about a proton (subject to the attractive Coulomb force $- (e^2/4\pi\varepsilon_o r^2)\hat{r}$) in orbits which satisfy the condition:

$$2\pi r = n\ \lambda_{\text{De Broglie}}, \qquad n = 1, 2, \ldots .$$

Since $\lambda_{\text{De Broglie}} = h/p$ this reduces to $p2\pi r = nh$, which may be generalized to the Wilson-Sommerfeld quantization condition

$$\oint p\ dq = nh \qquad n = 1, 2 \ldots \quad (1.1)$$

where $\oint$ implies a complete cycle, and p and q are conjugate variables. Equation (1.1) gives the correct quantized energies for the hydrogen atom 'by construction'. But it also gives the correct energy spectrum for a particle in a box with infinite walls

$$V(x) = 0 \qquad 0 < x < a$$
$$= \infty \qquad x < 0, \quad x > a.$$

EXAMPLE 1.1. Find the energy levels for a particle in a box with infinite walls:

$$V(x) = \infty \qquad x < 0, \quad x > a,$$

$$V(x) = 0 \qquad 0 < x < a.$$

In the region $0 < x < a$, $E = p^2/2m$.
Hence Equation (1.1) in this case becomes

$$\oint \sqrt{2mE}\ dx = nh$$

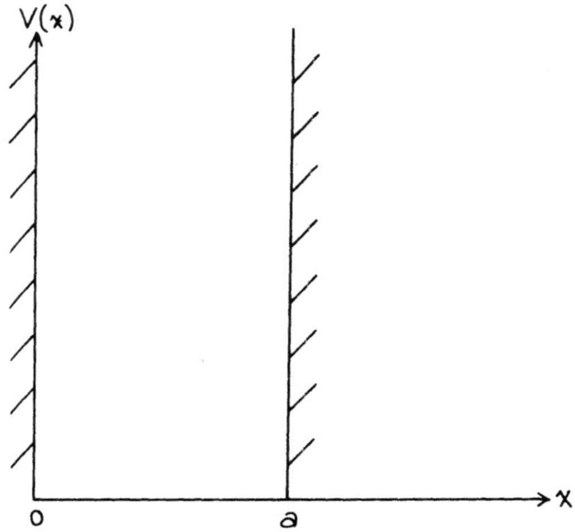

Figure 1.1. Potential in Example 1.1.

where a cycle involves x varying from $0 \to a$ and from $a \to 0$.

Integrating one obtains

$$2a\sqrt{2mE} = nh \quad \text{or} \quad E = \frac{n^2h^2}{8ma^2}, \quad n = 1, 2, \ldots . \tag{1.2}$$

By contrast the quantum-mechanical treatment of this problem involves solving the Schrödinger equation:

$$\left[-\frac{\hbar^2}{2m}\frac{d^2}{dx^2} + V(x)\right]\psi(x) = E\psi(x)\ldots \tag{1.3}$$

for $0 < x < a$ with boundary conditions $\psi(0)$, $\psi(a) = 0$, $\psi(x)$ being zero for $x < 0$, $x > a$.

The properly normalized eigenfunctions of Equation (1.3) which satisfy these boundary conditions are

$$\psi(x) = \sqrt{\frac{2}{a}}\sin kx \quad \text{with} \quad ka = n\pi, \quad \text{where} \quad \frac{\hbar^2k^2}{2m} = E.$$

Thus

$$\psi(0) = \sqrt{\frac{2}{a}}\sin 0 = 0$$

and

$$\psi(a) = \sqrt{\frac{2}{a}}\sin ka = 0 \quad \text{if} \quad ka = n\pi,\ n = 1, 2, \ldots,$$

while

$$\int_0^a |\psi(x)|^2 \, dx = 1$$

by construction.

Since

$$E = \frac{\hbar^2 k^2}{2m}$$

this implies

$$E = \frac{\hbar^2 n^2 \pi^2}{2ma^2} = \frac{h^2 n^2}{8ma^2} \quad n = 1, 2, \ldots,$$

exactly the result (1.2).

One can gain a little more insight as to the range of applicability of the Wilson-Sommerfeld quantization condition by studying slightly more complicated systems.

EXAMPLE 1.2. Find the energy levels for a particle in the potential:

$$\begin{aligned} V(x) &= 0 \qquad 0 < x < a, \\ &= V_0 \qquad a < x < a + b \\ &= \infty \qquad x < 0,\ x > a + b. \end{aligned}$$

(Assuming $E > V_0$ which is the interesting case.)

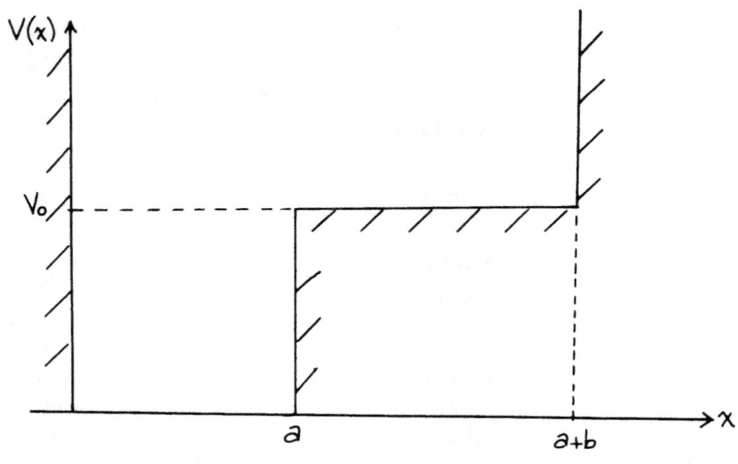

Figure 1.2. Potential in Example 1.2.

The Wilson-Sommerfeld quantization condition can be immediately applied to this case:

$$2\int_0^a p_1 dx + 2\int_a^{a+b} p_2\, dx = nh$$

where

$$p_1 = \hbar k_1 = \sqrt{2mE}\ , \qquad p_2 = \hbar k_2 = \sqrt{2m(E-V_0)}$$

i.e.

$$2\hbar(k_1 a + k_2 b) = nh$$

$$k_1 a + k_2 b = n\pi, \quad n = 1, 2, \ldots . \tag{1.4}$$

On the other hand solving the Schrödinger equation in the regions $0 < x < a$ and $a < x < a + b$ ($\psi(x)$ being zero for $x \le 0$, $x \ge a + b$) yields:

$$\psi(x) = A \sin k_1 x \qquad 0 < x < a$$

$$\psi(x) = B \sin k_2(x - a - b) \qquad a < x < a + b,$$

since $\psi(0)$ and $\psi(a+b)$ must be zero.

The continuity of $\psi(x)$, $d\psi(x)/dx$ at $x = a$ then implies

$$\tan k_1 a = -(k_1/k_2) \tan k_2 b \ldots \tag{1.5}$$

One notes that conditions (1.4) and (1.5) are different. Only if $k_2 \approx k_1$ i.e. $E >> V_0$ that is the total energy is large compared to the potential energy does Equation (1.5) reduce to Equation (1.4) since then $\tan k_1 a \approx -\tan k_2 b$ which is satisfied if

$$k_1 a + k_2 b = n\pi, \qquad n = 1, 2, \ldots .$$

Though the Wilson-Sommerfeld quantization condition was superseded by Quantum Mechanics (with the Schrödinger and Heisenberg formulations in the early twenties) as a calculational aid it has the advantage over the Schrödinger equation for instance that it is easier to work with since it involves an integral rather than a differential equation. However, it generally gives results which are reasonably accurate (i.e. in agreement with Quantum Mechanics) only when the energy is large compared to the potential under consideration.

If $V(x) = A|x|^p$ one can obtain the form of the energy sequence according to the Wilson-Sommerfeld quantization condition (1.1) as follows:

$$\oint \sqrt{2m} \sqrt{E_n - A|x|^p} \, dx = nh$$

can be written (for $E_n > 0$) as

$$\sqrt{2mE_n} \oint \sqrt{1-|u|^p} \, d\left(\frac{E_n}{A}\right)^{1/p} u = nh.$$

Hence

$$E_n^{\frac{p+2}{2p}} \left(\frac{\sqrt{2m}}{A^{1/p}}\right) \oint \sqrt{1 - |u|^p} \, du = nh$$

or generally:

$$|E_n| = n^{2p/(p+2)} \left(\frac{|A|^{1/p} h}{\sqrt{2m}\, I(p)}\right)^{2p/(p+2)} \qquad n = 1, 2, \ldots \quad (p > -2) \qquad (1.6)$$

where

$$I(p) = \oint \sqrt{1 - |u|^p} \, du \quad \text{if} \quad E_n > 0$$

and

$$I(p) = \oint \sqrt{|u|^p - 1} \, du \quad \text{if} \quad E_n < 0.$$

As $p \to \infty$, E_n in Equation (1.6) becomes $\propto n^2$, the result (1.2) for a particle in a box with infinite walls. [In detail $I(\infty) = 4$, $E_\infty = n^2 A^0 h^2/32m = n^2 h^2/32m$. This corresponds to $a = 2$ in Equation (1.2), i.e. $V = 0$ $|x| < 1$, $V = \infty$ $|x| > 1$.]

In several cases $I(p)$ can be easily evaluated.

EXAMPLE 1.3. Find the energy levels for a particle in the well $V(x) = Ax^2$, $-\infty < x < \infty$ using the Wilson-Sommerfeld quantization condition. If $p = 2$ Equation (1.6) reduces to

$$E_n = n\left(\frac{A}{2m}\right)^{1/2} \frac{h}{I(2)}$$

where

$$I(2) = 2\int_{-1}^{1} \sqrt{1-u^2} \, du = \pi$$

i.e.

$$E_n = n\hbar \sqrt{\frac{2A}{m}} \qquad n = 1, 2, \ldots . \qquad (1.7)$$

As opposed to the familiar Schrödinger equation result,

$$E_n = (n - \tfrac{1}{2})\, \hbar \sqrt{\frac{2A}{m}}, \quad n = 1, 2, \ldots . \tag{1.8}$$

EXAMPLE 1.4. Find the energy levels for a particle in the well

$$V(x) = Ax^2, \quad x > 0, \quad V(x) = \infty, \quad x < 0.$$

This problem goes through just as in Example 1.3 except

$$I\ (2) = 2 \int_0^1 \sqrt{1 - u^2}\ du = \frac{\pi}{2}$$

and hence

$$E_n = 2n\hbar \sqrt{\frac{2A}{m}} \quad n = 1, 2, \ldots . \tag{1.9}$$

As opposed to the Schrödinger equation result:

$$E_n = (2n-\tfrac{1}{2})\, \hbar \sqrt{\frac{2A}{m}}, \quad n = 1, 2, \ldots . \tag{1.10}$$

EXAMPLE 1.5. Find the energy levels for a particle in the well $V(x) = A|x|$ all x $(A > 0)$. Here $p = 1$ and Equation (1.6) reduces to:

$$E_n = n^{2/3} \left(\frac{Ah}{\sqrt{2m}\ I(1)}\right)^{2/3}$$

where

$$I(1) = 4 \int_0^1 \sqrt{1-x}\ dx = \frac{8}{3}$$

i.e.

$$E_n = n^{2/3} \left(\frac{A^2\hbar^2}{m}\right)^{1/3} \left(\frac{3\pi}{4\sqrt{2}}\right)^{2/3} \quad n = 1, 2, \ldots \tag{1.11}$$

as compared to the solution of the Schrödinger equation (see Equation (3.31)) for this problem in the limit of large E_n namely:

$$E_n = (n-\tfrac{1}{2})^{2/3} \left(\frac{A^2\hbar^2}{m}\right)^{1/3} \left(\frac{3\pi}{4\sqrt{2}}\right)^{2/3} \quad n = 1, 2, \ldots . \tag{3.31}$$

EXAMPLE 1.6. Find the energy levels for a particle in the well

$$V(x) = Ax \quad x > 0 \quad (A > 0).$$

This problem goes through as in Example 1.5 except

$$I(1) = 2\int_0^1 \sqrt{1 - x}\, dx = 4/3$$

and hence

$$E_n = n^{2/3}\left(\frac{A^2\hbar^2}{m}\right)^{1/3}\left(\frac{3\pi}{2\sqrt{2}}\right)^{2/3} \qquad (1.12)$$

as compared to the solution of the Schrödinger equation (see Equation (3.18)) for this problem in the limit of large E_n namely:

$$E_n = (n-\tfrac{1}{4})^{2/3}\left(\frac{A^2\hbar^2}{m}\right)^{1/3}\left(\frac{3\pi}{2\sqrt{2}}\right)^{2/3} \qquad (3.18)$$

EXAMPLE 1.7. Find the energy levels for a particle in the well $V(x) = -|A|/|x|$ all x. Here $p = -1$ and $E < 0$. Thus

$$|E_n| = n^{-2}\left(\frac{|A|^{-1}h}{\sqrt{2m}\, I(-1)}\right)^{-2}$$

where

$$I(-1) = 4\int_0^1 \sqrt{\frac{1}{u} - 1}\; du = 2\pi,$$

i.e.

$$E_n = -\frac{1}{n^2}\,\frac{2m|A|^2}{\hbar^2} \qquad (1.13)$$

Which is identical to the energy levels obtained for this problem using the Schrödinger equation (see Equation (3.32)).

EXAMPLE 1.8. Find the energy levels for a particle in the well

$$V(x) = -\frac{|A|}{x}, \quad x > 0 \quad V(x) = \infty, \quad x < 0.$$

One proceeds as in Example 1.7 except

$$I\,(-1) = 2\int_0^1 \sqrt{\frac{1}{u} - 1}\, du = \pi$$

and hence

$$E_n = -\frac{m|A|^2}{n^2 2\hbar^2} \quad n = 1, 2, \ldots \qquad (1.14)$$

which is identical to the energy levels obtained for this problem using the Schrödinger equation (see Equation (3.23)).

EXAMPLE 1.9. Find the energy levels for a particle in the well

$$V(x) = A|x|^{1/2} \quad \text{all} \quad x.$$

Here

$$p = \tfrac{1}{2} \quad \text{and} \quad E > 0.$$

Thus

$$E_n = n^{2/5}\left(\frac{A^2 h}{\sqrt{2m}\, I(\frac{1}{2})}\right)^{2/5}$$

where

$$I(\tfrac{1}{2}) = 4\int_0^1 \sqrt{1 - u^{\frac{1}{2}}}\, du = \frac{32}{15}$$

i.e.

$$E_n = n^{2/5}\left(\frac{A^2 h 15}{\sqrt{2m}\; 32}\right)^{2/5} \quad \ldots \; n = 1, 2 \qquad (1.15)$$

EXAMPLE 1.10. Find the energy levels for a particle in the well

$$V(x) = Ax^{\frac{1}{2}} \quad x > 0, \quad V(x) = \infty \quad x < 0.$$

One proceeds as in Example 1.9 except

$$I(\tfrac{1}{2}) = 2\int_0^1 \sqrt{1 - u^{\frac{1}{2}}}\, du = \frac{16}{15}$$

i.e.

$$E_n = n^{2/5}\left(\frac{A^2 h 15}{\sqrt{2m}\; 16}\right)^{2/5} \quad n = 1, 2, \ldots . \qquad (1.16)$$

Using this technique one may also obtain analytic expressions for E_n if $V(x) = Ax^{\frac{1}{4}}$ etc. As $p \to 0$ the levels go as n^p i.e. become very close together or one approaches a continuum situation.

EXAMPLE 1.11. Find the energy levels for a particle in the well

$$V(x) = -|A|\,|x|^{-\frac{1}{2}} \quad \text{all} \quad x.$$

Here $p = -\frac{1}{2}$ and $E < 0$. Thus

$$|E_n| = n^{-2/3}\left(\frac{|A|^{-2}\,h}{\sqrt{2m}\;I(-\frac{1}{2})}\right)^{-2/3} \qquad n = 1, 2, \ldots$$

where

$$I(-\tfrac{1}{2}) = 4\int_0^1 \sqrt{u^{-\frac{1}{2}} - 1}\,du = \pi$$

$$E_n = -n^{-2/3}\left(\frac{|A|^{-2}h}{\sqrt{2m}\;\pi}\right)^{-2/3} . \tag{1.17}$$

EXAMPLE 1.12. Find the energy levels for a particle in the well

$$V(x) = -|A|x^{-\frac{1}{2}}, \quad x > 0, \quad V(x) = \infty, \quad x < 0, \quad E < 0.$$

One proceeds as in Example 1.11 except

$$I(-\tfrac{1}{2}) = 2\int_0^1 \sqrt{u^{-\frac{1}{2}} - 1}\;du = \frac{\pi}{2}$$

i.e.

$$E_n = -\,n^{-2/3}\left(\frac{|A|^{-2}\;2h}{\sqrt{2m}\;\pi}\right)^{-2/3} . \tag{1.18}$$

The results of Examples 1.3 - 1.12 are summarized in Table 1.1.

For two-dimensional systems (or three-dimensional systems where a particle moves in a plane chosen for convenience to be the x-y plane), and the potential only depends on ρ,

$$E = \frac{p_\rho^2}{2m} + \frac{p_\rho^2}{2m\rho^2} + V(\rho). \tag{1.19}$$

Hence

$$\oint p_\phi d_\phi = n_\phi h \qquad n_\phi = 1, 2 \ldots \tag{1.20}$$

which implies

$$p_\phi = n_\phi \hbar$$

and

$$\oint p_\rho d_\rho = n_\rho h \qquad n_\rho = 1, 2, \ldots \tag{1.21}$$

i.e.

TABLE 1.1
Energy levels for various potentials $V(x) = A|x|^p$

P	Range	Wilson-Sommerfeld quantization condition	Schrödinger result
1	$x \geq 0$	$E_n = n^{2/3}\left(\frac{A^2\hbar^2}{m}\right)^{1/3}\left(\frac{3\pi}{2\sqrt{2}}\right)^{2/3}$	(for large E) $E_n = \left(n-\frac{1}{4}\right)^{2/3}\left(\frac{A^2\hbar^2}{m}\right)^{1/3}\left(\frac{3\pi}{2\sqrt{2}}\right)^{2/3}$
1	$-\infty<x<\infty$	$E_n = n^{2/3}\left(\frac{A^2\hbar^2}{m}\right)^{1/3}\left(\frac{3\pi}{4\sqrt{2}}\right)^{2/3}$	(for large E) $E_n = \left(n-\frac{1}{2}\right)^{2/3}\left(\frac{A^2\hbar^2}{m}\right)^{1/3}\left(\frac{3\pi}{4\sqrt{2}}\right)^{2/3}$
2	$x \geq 0$	$E_n = 2n\left(\frac{A}{2m}\right)^{1/2}\frac{h}{\pi}$	$E_n = \left(2n-\frac{1}{2}\right)\left(\frac{A}{2m}\right)^{1/2}\frac{h}{\pi}$
2	$-\infty<x<\infty$	$E_n = n\left(\frac{A}{2m}\right)^{1/2}\frac{h}{\pi}$	$E_n = \left(n-\frac{1}{2}\right)\left(\frac{A}{2m}\right)^{1/2}\frac{h}{\pi}$
-1	$x \geq 0$	$E_n = -\frac{m\|A\|^2}{n^2 2\hbar^2}$	Same
-1	$-\infty<x<\infty$	$E_n = -\frac{2m\|A\|^2}{n^2\hbar^2}$	Same
$\frac{1}{2}$	$x \geq 0$	$E_n = n^{2/5}\left(\frac{A^2 h 15}{\sqrt{2m}\ 16}\right)^{2/5}$	
$\frac{1}{2}$	$-\infty<x<\infty$	$E_n = n^{2/5}\left(\frac{A^2 h 15}{\sqrt{2m}\ 32}\right)^{2/5}$	
$-\frac{1}{2}$	$x \geq 0$	$E_n = -n^{-2/3}\left(\frac{\|A\|^{-2} 2h}{\sqrt{2m}\ \pi}\right)^{-2/3}$	
$-\frac{1}{2}$	$-\infty<x<\infty$	$E_n = -n^{-2/3}\left(\frac{\|A\|^{-2} h}{\sqrt{2m}\ \pi}\right)^{-2/3}$	

$$\oint \sqrt{2mE - \frac{(n_\phi \hbar)^2}{\rho^2} - 2mV(\rho)}\ d\rho = n_\rho h \qquad n_\rho = 1, 2, \ldots . \tag{1.22}$$

The integral in Equation (1.22) can be evaluated analytically for certain problems.

EXAMPLE 1.11. Find the energy levels for the potential $V(\rho) = \frac{1}{2}m\omega^2\rho^2$. From Equation (1.22) one has

$$\oint \sqrt{2mE\rho^2 - (n_\phi\hbar)^2 - m^2\omega^2\rho^4}\ \frac{d\rho}{\rho} = n_\rho h \tag{1.23}$$

One can either evaluate integral (1.23) using complex integration[1]) or by elementary methods.

Thus defining $A \equiv -(n_\phi\hbar)^2$, $B \equiv 2mE$, $C \equiv -m^2\omega^2$ $u = \rho^2$

$$\oint \sqrt{A + Bu + Cu^2}\ \frac{du}{u} = 2n_\rho h$$

$$= 2\left(A\int_{u_{min}}^{u_{max}} \frac{du}{u\sqrt{A+Bu+Cu^2}} + \frac{B}{2}\int_{u_{min}}^{u_{max}} \frac{du}{\sqrt{A+Bu+Cu^2}} + \frac{1}{2}\int_{u_{min}}^{u_{max}} \frac{d(A+Bu+Cu^2)}{\sqrt{A+Bu+Cu^2}}\right) =$$

$$2\left(A\frac{1}{\sqrt{-A}}\sin^{-1}\frac{Bu+2A}{u\sqrt{B^2-4AC}} + \frac{B}{2}\frac{1}{\sqrt{-C}}\sin^{-1}\left(\frac{-2Cu-B}{\sqrt{B^2-4AC}}\right) + \sqrt{A+Bu+Cu^2}\,\Bigg|_{u_{min}}^{u_{max}}\right) = 2n_c h$$

where u_{max} and u_{min} are determined by requiring that the integrand $\frac{\sqrt{A+Bu+Cu^2}}{u}$ vanish i.e. $u_{\substack{max\\min}} = \frac{-B\mp\sqrt{B^2-4AC}}{2C}$ and $B^2 > 4AC$

i.e. $E \geq n_\phi\hbar\,\omega$.

This yields:

$$(-n_\phi\hbar)\pi + \frac{E\pi}{\omega} = n_\rho h$$

or

$$E = (2n_\rho + n_\phi)\hbar\omega \qquad (1.24)$$

where

$$\left.\begin{matrix} n_\rho \\ n_\rho \end{matrix}\right\} 1,\ 2,\ \ldots$$

as opposed to the standard Schrödinger result for this system (see chapters 8 and 12)

$$E = (2n + m - 2)\hbar\omega \qquad \left.\begin{matrix} n \\ m \end{matrix}\right\} = 1,\ 2\ldots \qquad (8.14)$$

Besides its shortcoming that it gives quantitized energies which are as a rule only approximately correct for energies large compared to the potentials involved, the Wilson-Sommerfeld procedure says nothing about the evaluation of probability distributions, transition rates etc. for which there are standard techniques in quantum mechanics.

On a more positive note the quantization condition Equation (1.1) is also a consequence of applying the W.K.B. approximation to the Schrödinger equation[2]) with the modification that nh must be replaced by $(n + \frac{1}{2})h$ to get the W.K.B. approximation result.

References

1. H. Goldstein, Classical Mechanics, Addison Wesley (1950), p. 300.
2. E. Merzbacher, Quantum Mechanics, Wiley (1970), p. 123.

CHAPTER 2

The Delta Function, Completeness and Closure

The delta function is defined to have the following properties (in one dimension)

$$\delta(x-x') = 0 \quad x \neq x'$$
$$\int \delta(x-x')\, dx = 1 \tag{2.1}$$

These have as a consequence that

$$\int_{-\infty}^{+\infty} f(x)\delta(x-x')\, dx = f(x'). \tag{2.2}$$

One way to get some insight into this useful function and expressions for it in terms of standard functions is to use 'the principle of completeness'. The principle of completeness allows one to expand an arbitrary function in terms of any complete orthonormal set. Thus if $\psi(x)$ is an arbitrary function

$$\psi(x) = \sum_n a_n \phi_n(x), \tag{2.3}$$

if the complete set $\phi_n(x)$ chosen is a discrete set, or

$$\phi(x) = \int a(k)\ \phi_k(x)\ dk \tag{2.4}$$

if the complete set $\phi_k(x)$ involves continuous functions, where

$$a_n = \int \phi_n^*(x)\psi(x)\ dx,\ a(k) = \int \phi_k^*(x)\psi(x)\ dx. \tag{2.5}$$

Expanding the delta function $\delta(x-x')$ in terms of a complete set of <u>discrete</u> functions implies:

$$\delta(x-x') = \sum_n a_n \phi_n(x)$$

where

$$a_n = \int \phi_n^*(x)\delta(x-x')\, dx = \phi_n^*(x')$$

i.e.

$$\delta(x-x') = \sum_{n=0}^{\infty} \phi_n^*(x')\phi_n(x), \tag{2.6}$$

and similarly expanding the delta function in terms of a complete set of continuous functions implies

$$\delta(x-x') = \int_{-\infty}^{+\infty} \phi_k^*(x')\phi_k(x)\, dk. \tag{2.7}$$

EXAMPLE 2.1. Suppose one uses as a complete discrete set the eigenfunctions of a particle in an infinite square well potential (a box with infinite walls)

$$V(x) = 0 \quad -\frac{a}{2} < x < \frac{a}{2}$$

$$V(x) = \infty \quad -\frac{a}{2} > x,\ x > \frac{a}{2},$$

for the expansion (2.6) with the choice $x' = 0$. The normalized even subset of the above eigenfunctions (the rest, i.e. the odd subset is zero at $x = 0$ and does not contribute to the integral (2.5) for a_n, and hence to the sum (2.6)) is

$$\phi_n(x) = \sqrt{\frac{2}{a}} \cos \frac{(2n+1)}{a} \pi x \quad n = 0, 1\ldots \quad -\frac{a}{2} < x < \frac{a}{2}. \tag{2.8}$$

$$\phi_n(x) = 0 \qquad x < -\frac{a}{2}, \quad x > \frac{a}{2}.$$

Hence a possible representation of the delta function is

$$\delta(x) = \frac{2}{a} \sum_{n=0}^{\infty} \cos \frac{(2n+1)\pi x}{a} \quad -\frac{a}{2} < x < \frac{a}{2} \tag{2.9}$$

$$= 0 \qquad x < -\frac{a}{2} \quad x > \frac{a}{2}.$$

EXAMPLE 2.2. Show Equation (2.9) is consistent with Equation (2.1). One notes, interchanging integration and summation that

$$\int \frac{2}{a} \sum_{n=0}^{\infty} \cos \frac{2n+1}{a} \pi x \, dx = \frac{2}{a} \sum_{n=0}^{\infty} \int_{-a/2}^{a/2} \cos \frac{(2n+1)\pi x}{x} \, dx =$$

$$= \frac{4}{\pi} \sum_{n=0}^{\infty} \frac{(-1)^n}{2n+1} = 1,$$

consistent with the integral in expression (2.1).

Considering the first two terms in expansion (2.9) as a crude approximation one gets the approximate representation

$$\delta(x) \approx \frac{2}{a}\left(\cos \frac{\pi x}{a} + \cos \frac{3\pi x}{a} \right) = \frac{4}{a} \cos \frac{\pi x}{a} \cos \frac{2\pi x}{a} , \qquad (2.10)$$

plotted in Figure 2.1.

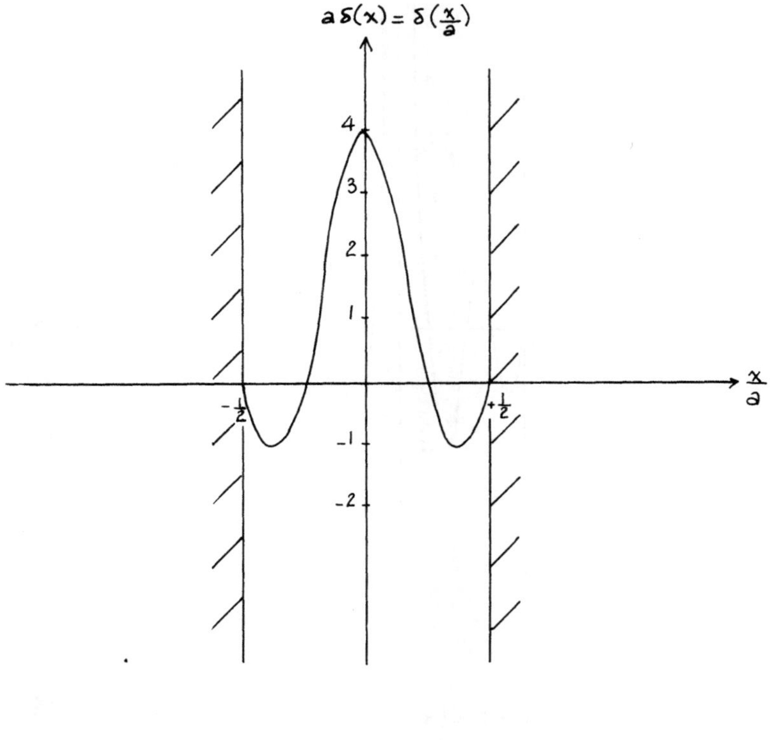

Figure 2.1. Plot of $\delta(x) \approx \frac{4}{a} \cos \frac{\pi x}{a} \cos \frac{2\pi x}{a}$.

Considering the first four terms in expansion (2.9) one gets a somewhat better approximate representation of the delta function

$$\delta(x)\ \frac{2}{a}\left(\cos\frac{\pi x}{a} + \cos\frac{3\pi x}{a} + \cos\frac{5\pi x}{a} + \cos\frac{7\pi x}{a}\right) =$$

$$\frac{8}{a}\cos\frac{\pi x}{a}\cos\frac{2\pi x}{a}\cos\frac{4\pi x}{a}\ . \qquad (2.11)$$

This is plotted in Figure 2.2.

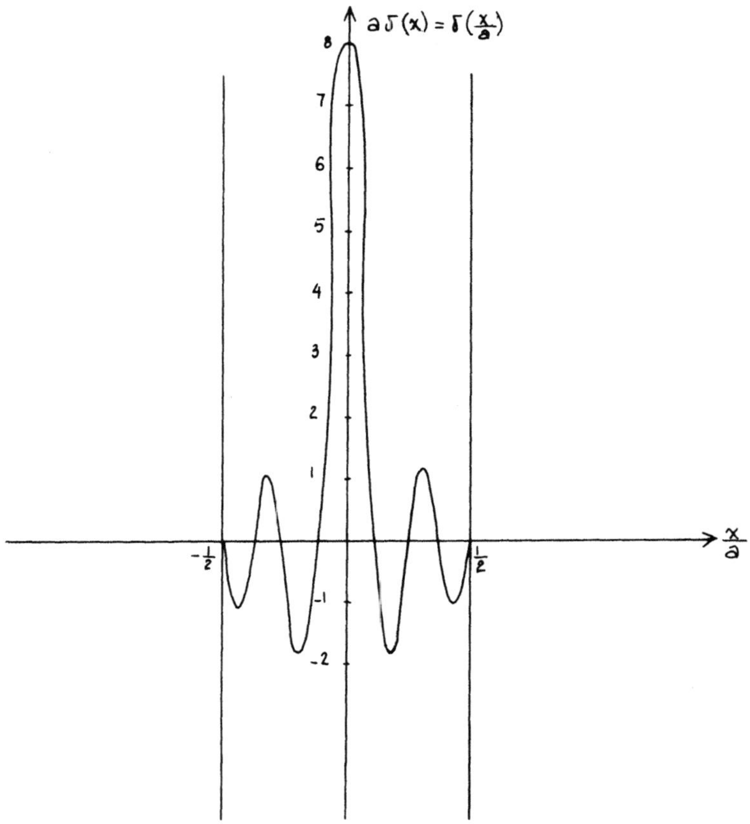

Figure 2.2. Plot of $\delta(x) \simeq \frac{8}{a}\cos\frac{\pi x}{a}\cos\frac{2\pi x}{a}\cos\frac{4\pi x}{a}$.

One notes that the central maximum (about x = 0) gets progressively sharper and the secondary maxima less important as the number of terms increases.

Expression (2.9) resembles a Fourier expansion with the constraint the expansion is zero at $x = \pm a/2$ and only valid for $-a/2 < x < a/2$.

EXAMPLE 2.3. Consider instead for the complete, discrete set in the expansion of the delta function, the eigenfunctions of a particle in a harmonic oscillator potential:

$$V(x) = \frac{\hbar}{2m}\frac{x^2}{b^4} .$$

The eigenfunctions of this potential which are non zero at x = 0 are

$$\phi_n(x) = \frac{1}{2^{n/2}}\left(\frac{1}{n!}\right)^{\frac{1}{2}}\left(\frac{1}{\pi b^2}\right)^{\frac{1}{4}} e^{-x^2/2b^2} H_n\left(\frac{x}{b}\right) \text{ (n even)}. \tag{2.12}$$

Hence an alternative representation of the δ function is

$$\delta(x) = \frac{1}{\sqrt{\pi}b} e^{-x^2/2b^2} \sum_{n=0}^{\infty} \frac{1}{2^n n!} H_n(\tfrac{x}{b})H_n(0). \tag{2.13}$$

where one need not specify the sum is only over even n since $H_n(0)$ is zero for odd n

The mathematical identity (known as Mehler's formula[1])

$$\frac{1}{\sqrt{1-t^2}} \exp\left(\frac{4xyt-(x^2+y^2)(1+t^2)}{2(1-t^2)}\right) = \exp -\left(\frac{x^2+y^2}{2}\right) \sum_{n=0}^{a} \frac{H_n(x)H_n(y)t^n}{2^n n!} \tag{2.14}$$

with the substitution $y = 0$, $x \to x/b$ becomes

$$\frac{1}{\sqrt{1-t^2}} \exp\left(-\frac{x^2(1+t^2)}{2b^2(1-t^2)}\right) = \exp\left(\frac{-x^2}{2b^2}\right) \sum_{n=0}^{\infty} \frac{H_n(\frac{x}{b})H_n(0)t^n}{2^n n!} .$$

Substituting this expression in Equation (2.13) yields

$$\delta(x) = \frac{1}{\sqrt{\pi}} \lim_{t\to 1} \frac{1}{b\sqrt{1-t^2}} \exp \frac{-x^2(1+t^2)}{2b^2(1-t^2)}$$

or in other words provided one makes the substitution $\varepsilon = b^2(1-t^2)$,

$$\delta(x) = \frac{1}{\sqrt{\pi}} \lim_{\varepsilon\to 0} \frac{1}{\sqrt{\varepsilon}} \exp -\frac{x^2}{\varepsilon}. \tag{2.15}$$

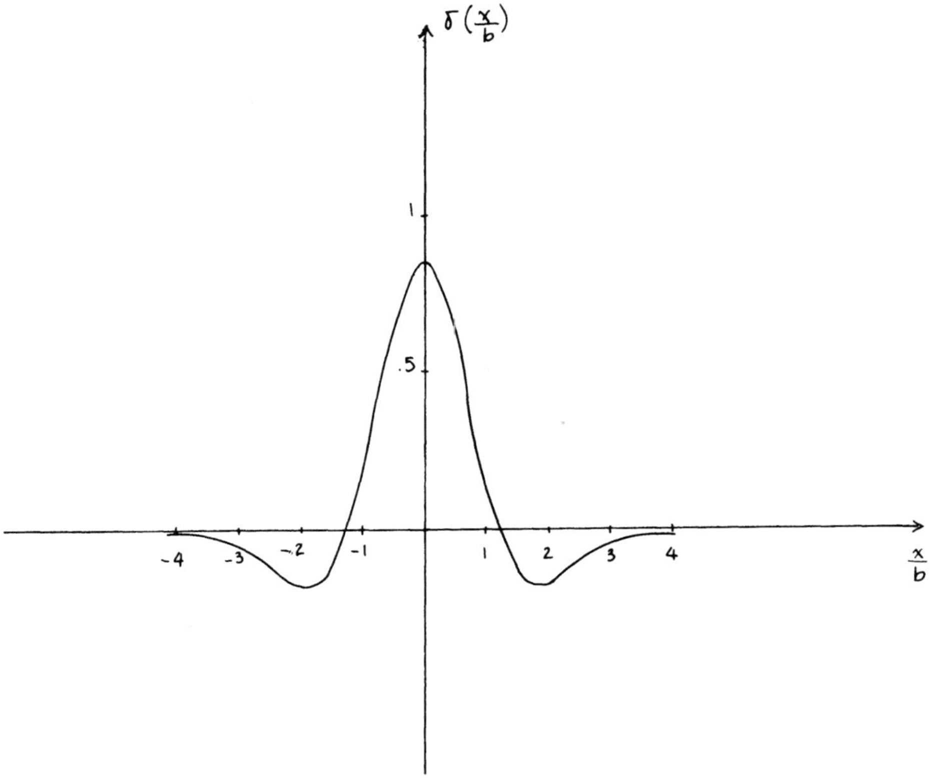

Fig. 2.3. Plot of $\delta(\frac{x}{b}) = b\delta(x) \approx \frac{3}{2}\frac{e^{-x^2/2b^2}}{\sqrt{\pi}}\left(1-\frac{2}{3}\frac{x^2}{b^2}\right)$.

Expression (2.15) is a standard representation of the delta function in terms of a limit.

In Figures 2.3 and 2.4 are plotted two approximate expressions for the delta function using Equation (2.13). The first involves including the first two non-zero terms and the second the first three non-zero terms of this expression,

$$\delta(x) \approx \frac{3}{2}\frac{e^{-x^2/2b^2}}{\sqrt{\pi}\ b}\left(1 - \frac{2}{3}\frac{x^2}{b^2}\right),$$

$$\delta(x) \approx \frac{15}{8\sqrt{\pi}\ b}e^{-x^2/2b^2}\left(1 - \frac{4x^2}{3b^2} + \frac{4}{15}\frac{x^4}{b^4}\right), \tag{2.16}$$

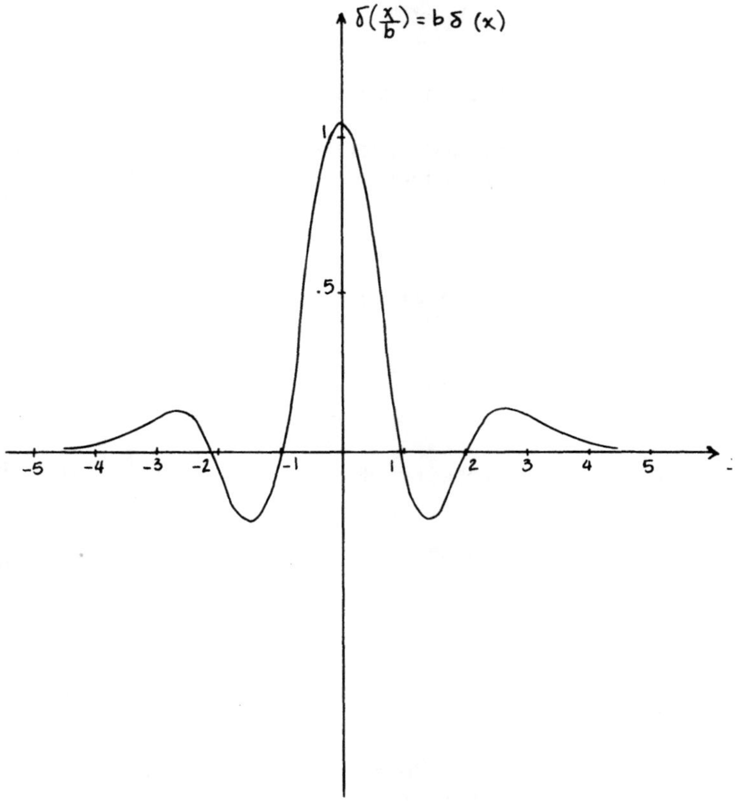

Figure 2.4. Plot of $b\delta(x) \approx \frac{15x}{8\sqrt{\pi}} e^{-x^2/2b^2}\left(1 - \frac{4}{3}\frac{x^2}{b^2} + \frac{4}{15}\frac{x^4}{b^4}\right)$.

respectively. The general features and trends are similar to those of Figure 2.1 and 2.2.

EXAMPLE 2.4. Show Equation (2.13) is consistent with Equation (2.2). One notes, interchanging integration and summation that since $H_0(\xi) = 1$

$$\int_{-\infty}^{+\infty} dx e^{-x^2/2b^2}\delta(x) = \frac{1}{\sqrt{\pi}\, b}\int_{-\infty}^{+\infty} e^{-x^2/b^2} \sum_{n=0}^{\infty} \frac{1}{2^n n!} H_n\left(\frac{x}{b}\right) H_n(0)\, dx =$$

$$\frac{1}{\sqrt{\pi}\, b} \sum_{n=0}^{\infty} \frac{1}{2^n n!} H_n(0) \int_{-\infty}^{+\infty} e^{-x^2/b^2} H_n\left(\frac{x}{b}\right) H_0\left(\frac{x}{b}\right) dx =$$

$$\frac{1}{\sqrt{\pi}} \sum_{n=0}^{\infty} \frac{1}{2^n n!} H_n(0)\delta_{n0} \sqrt{\pi} = 1,$$

consistent with the integral in expression (2.2).

EXAMPLE 2.5. Use as a continuous representation of the delta function the set of free particle wavefunctions ($V(x) = 0$, all x)

$$\phi_k(x) = Ae^{ikx}.$$

This implies using Equation (2.4), (2.5)

$$\delta(x) = A^2 \int_{k=-\infty}^{k=\infty} e^{ikx}\, dk = A^2 \lim_{k\to\infty} \frac{e^{ikx}-e^{-ikx}}{ix} = 2A^2 \lim_{k\to\infty} \frac{\sin kx}{x}.$$

One can obtain using the property of Equation (2.1) $\int_{-\infty}^{+\infty} \delta(x)\, dx = 1$, that

$$1 = 2A^2 \lim_{k\to\infty} \int_{-\infty}^{+\infty} \frac{\sin kx}{x}\, dx = 2A^2\pi \text{ (a result independent of k).}$$

Hence

$$\delta(x) = \frac{1}{\pi} \lim_{k\to\infty} \frac{\sin kx}{x} \tag{2.17}$$

is a second expression for the delta function in terms of a limit. Using the above value of A the set of free particle wavefunctions with 'delta function' normalization becomes

$$\phi_k(x) = \frac{1}{\sqrt{2\pi}} e^{ikx}, \tag{2.18}$$

while $\delta(x)$ can also be written

$$\delta(x) = \frac{1}{2\pi} \int_{-\infty}^{+\infty} e^{ikx}\, dk. \tag{2.19}$$

The three dimensional delta function $\delta(\vec{r} - \vec{r}')$ in spherical coordinates can be written as follows

$$\delta(\vec{r}-\vec{r}') = \frac{\delta(r-r')}{r^2} \delta(\cos\theta - \cos\theta')\delta(\phi-\phi') =$$

$$= \sum_{n,\ell,m} R_{n\ell}(r)R_{n\ell}(r')Y_m^{*\ell}(\theta,\phi)\, Y_m^{\ell}(\theta',\phi'). \tag{2.20}$$

If one multiples both sides of this equation by $Y_p^q(\theta, \phi)Y_p^{*q}(\theta', \phi')$ and integrates over $d\Omega d\Omega'$ one obtains

$$\frac{\delta(r-r')}{r^2} = \sum_n R_{nq}(r)R_{nq}(r'),$$

and for three dimensional systems quite generally one can thus write for the delta function (for any choice of orbital quantum number ℓ):

$$\delta(r-r') = \sum_{n=0}^{\infty} u^*_{n\ell}(r')u_{n\ell}(r) \tag{2.21a}$$

where $u_{n\ell}(r)$ is the solution, subject to the condition $u_{n\ell}(0) = 0$ of the radial Schrödinger equation:

$$\left\{-\frac{\hbar^2}{2m}\frac{d^2}{dr^2} + V(r) + \frac{\hbar^2\ell(\ell+1)}{2mr^2} - E\right\} u_{n\ell}(r) = 0 \quad (r \geq 0),$$

i.e. $u_{n\ell}(r) = rR_{n\ell}(r)$.

The choice of central potential $V(r)$ determines the detailed form of $u_{n\ell}(r)$. The corresponding one dimensional expression is Equation (2.6) where x extends over all space, and $\Phi_n(x)$ is the solution of the one dimensional Schrödinger equation:

$$\left\{-\frac{\hbar^2}{2m}\frac{d^2}{dx^2} + V(x) - E\right\}\phi_n(x) = 0, \qquad (-\infty < x < \infty),$$

where similarly the choice of $V(x)$ determines the detailed form of $\Phi_n(x)$. For continuous states Equation (2.21) is replaced by:

$$\delta(r-r') = \int u^*_{k\ell}(r')\, u_{k\ell}(r)\, dk, \tag{2.21b}$$

analogous to the one dimensional Equation (2.7).

Equation (2.6), (2.7), (2.21a), and (2.21b) are illustrations of the 'closure' property of quantum mechanical wavefunctions.

EXAMPLE 2.6. Obtain an expression for $\delta(r-r')$, $|r-r'| > 0$ if $V(r)$ is the three dimensional simple harmonic oscillator potential

$$V(r) = \frac{\hbar^2 r^2}{2mb^4} \qquad r > 0$$
$$= 0 \qquad r < 0.$$

Inserting the detailed solutions $u_{n\ell}(r)$ for this potential (cf.) Equation 8.13) in expression (2.21) one obtains:

$$\delta(r-r') = \frac{2e^{-(r^2+r'^2)/2b^2}(rr')^{\ell+1}}{b^{2\ell+3}\Gamma^2(\ell+3/2)} \sum_{n=0}^{\infty} \frac{\Gamma(n+\ell+3/2)}{n!} {}_1F_1\left(-n;\ell+\frac{3}{2};\frac{r^2}{b^2}\right),$$

$${}_1F_1\left(-n;\ell+\frac{3}{2};\frac{r'^2}{b^2}\right), \tag{2.22}$$

independent of the value of b, ℓ where r, r' > 0.
If $\ell = 0$ for example

$$\delta(r-r') = \frac{8e^{-(r^2+r'^2)/2b^2}(rr')/b^2}{b\pi} \sum_{n=0}^{\infty} \frac{\Gamma(n+3/2)}{n!} {}_1F_1\left(-n;\frac{3}{2};\frac{r^2}{b^2}\right),$$

$${}_1F_1\left(-n;\frac{3}{2};\frac{r'^2}{b^2}\right), \tag{2.23}$$

Substituting Equation (2.12) the solutions of the one dimensional harmonic oscillator in Equation (2.6) one has analogously

$$\delta(x-x') = \frac{e^{-(x^2+x'^2)/2b^2}}{\sqrt{\pi}\, b} \sum_{n=0}^{\infty} \frac{1}{2^n n!} H_n\left(\frac{x}{b}\right)H_n\left(\frac{x'}{b}\right). \tag{2.24}$$

But

$$H_{2n+1}(\xi) = \frac{(-1)^n}{n!}(2n+1)!2\xi\, {}_1F_1\left(-n;\frac{3}{2};\xi^2\right)^{1}.$$

Substituting this expression into Equation (2.23) and using Legendre's duplication formula[2])

$$n!\Gamma(n+3/2) = \frac{(2n+1)!\sqrt{\pi}}{2^{2n+1}},$$

Equation (2.23) can be written in a form similar to Equation 2.24, namely

$$\delta(r-r') = \frac{2e^{-(r^2+r'^2)/2b^2}}{\sqrt{\pi}\, b} \sum_{\substack{n\,\text{odd}\\1,3\ldots}} \frac{H_n(\frac{r}{b})H_n(\frac{r'}{b})}{n!\,2^n} \tag{2.25}$$

Equation (2.25) involves a sum only over odd n terms for which $H_n(\xi)$ is zero if $\xi = 0$, since r or r' cannot be zero. Thus $\delta(r-r')$ in Equation (2.25) vanishes if either r or r' is zero. Also the additional factor 2 is there in Equation (2.25) because x, x' extend from $-\infty$ to $+\infty$ whereas r, r' extend from $0 \to \infty$. Using the Hille-Hardy formula[3])

$$(xy)^{\alpha/2} \frac{e^{-(x+y)/2}}{\Gamma^2(\alpha+1)} \sum_{n=0}^{\infty} \frac{\Gamma(n+\alpha+1)}{n!} {}_1F_1(-n;\alpha+1;x)\, {}_1F_1(-n;\alpha+1;\, y)t^n =$$

$$\frac{t^{-\alpha/2}}{1-t}\, e^{-\frac{1}{2}(x+y)\frac{1+t}{1-t}}\, I_\alpha\!\left(\frac{2(xyt)^{\frac{1}{2}}}{1-t}\right), \tag{2.26}$$

one can sum Equation (2.22). Here

$$I_\alpha(x) = i^{-\alpha} J_\alpha(ix) = \sqrt{\frac{2x}{\pi i^{2\alpha-1}}}\, j_{\alpha-\frac{1}{2}}(ix).$$

Thus,

$$\delta(r-r') = \lim_{t\to 1} \frac{2(rr')^{\frac{1}{2}}}{b^2} \frac{t^{-\frac{\ell+\frac{1}{2}}{2}}}{1-t}\, e^{-\frac{(r^2+r'^2)}{2b^2}\frac{1+t}{1-t}}\, I_{\ell+\frac{1}{2}}\!\left(\frac{2rr't^{\frac{1}{2}}}{b^2(1-t)}\right)$$

$$(r,\ r' > 0), \tag{2.27}$$

which can be written in terms of spherical Bessel functions:

$$\delta(r-r') = \lim_{t\to 1} \frac{4rr'}{i^\ell b^3\sqrt{\pi}} \frac{t^{-\ell/2}}{(1-t)^{3/2}}\, e^{-\frac{r^2+r'^2}{2b^2}\frac{1+t}{1-t}}\, j_\ell\!\left(\frac{i2rr't^{\frac{1}{2}}}{b^2(1-t)}\right)$$

$$(r,\ r' > 0). \tag{2.28}$$

If one makes the substitution $\varepsilon = (1-t)b^2$,

$$\delta(r-r') = \lim_{\varepsilon\to 0} \frac{4rr'}{i^\ell\sqrt{\pi}}\, e^{-\frac{r^2+r'^2}{\varepsilon}}\, j_\ell\left(\frac{i2rr'}{\varepsilon}\right) \qquad (r,\ r' > 0) \tag{2.29}$$

or

$$\delta(r-r') = \lim_{\varepsilon\to 0} \frac{2(rr')^{\frac{1}{2}}}{\varepsilon} \frac{e^{-\frac{r^2+r'^2}{\varepsilon}}}{\varepsilon^{3/2}}\, I_{\ell+\frac{1}{2}}\!\left(\frac{2rr'}{\varepsilon}\right) \qquad (r,\ r' > 0). \tag{2.30}$$

For $\ell = 0$, since

$$I_{\frac{1}{2}}(\xi) = (\tfrac{1}{2}\pi\xi)^{-\frac{1}{2}} \sinh \xi,^{4})$$

$$\delta(r-r') = \lim_{\varepsilon\to 0} \frac{2}{\sqrt{\pi}\sqrt{\varepsilon}}\, e^{-\frac{r^2+r'^2}{\varepsilon}} \sinh \frac{2rr'}{\varepsilon}. \tag{2.31}$$

But as $\varepsilon \to 0$ $\sinh \frac{2rr'}{\varepsilon} \to \frac{1}{2}e^{2rr'/\varepsilon}$ and Equation (2.31) reduces to

$$\delta(r-r') = \frac{1}{\sqrt{\pi}} \lim_{\varepsilon\to 0} \frac{2}{\sqrt{\varepsilon}} e^{-\frac{r^2+r'^2}{\varepsilon}} \frac{e^{\frac{2rr'}{\varepsilon}}}{2} = \frac{1}{\sqrt{\pi}} \lim_{\varepsilon\to 0} \frac{1}{\sqrt{\varepsilon}} e^{-\frac{(r-r')^2}{\varepsilon}} \tag{2.32}$$

an obvious generalization of Equation (2.15).

EXAMPLE 2.7. Verify Equation (2.25) is consistent with Equation (2.2). According to Equation (2.2)

$$\int \delta(r-r') H_m(\tfrac{r}{b})\, e^{-r^2/2b^2}\, dr = H_m(\tfrac{r'}{b})\, e^{-r'^2/2b^2}.$$

Multiplying Equation (2.25) by $H_m(\frac{r}{b})\, e^{-r^2/2b^2}$ (m_{odd}) and integrating both sides over r one obtains

$$H_m\left(\frac{r'}{b}\right) e^{-r'^2/2b^2} \stackrel{?}{=} \frac{2e^{-r'^2/2b^2}}{\sqrt{\pi}} \sum_{n_{odd}} \frac{H_n(\frac{r'}{b})}{n!2^n} \int_0^\infty H_n(\tfrac{r}{b}) H_m(\tfrac{r}{b}) e^{-r^2/b^2} d(\tfrac{r}{b}) \tag{2.33}$$

where integration and summation have been interchanged. But the integrand in Equation (2.33) is always even. Hence the rhs of Equation 2.33 can be written

$$\frac{e^{-r'^2/2b^2}}{\sqrt{\pi}} \sum_{n_{odd}} \frac{H_n(\frac{r'}{b})}{n!2^n} \int_{-\infty}^{+\infty} H_n(\tfrac{r}{b}) H_m(\tfrac{r}{b}) e^{-r^2/b^2} d(\tfrac{r}{b}) =$$

$$\frac{e^{-r'^2/2b^2}}{\sqrt{\pi}} \sum_{n_{odd}} \frac{H_n(\frac{r'}{b})}{n!2^n} \delta_{nm} m! 2^m \sqrt{\pi} = e^{-r'^2/2b^2} H_m(\tfrac{r'}{b}),$$

which is identical with the lhs of this equation.

Since completeness enables one to expand any arbitrary function in terms of a complete set it can also be used to describe what happens if the potential of a system suddenly changes without the wavefunction undergoing any modification:

EXAMPLE 2.8. Suppose a particle is in the ground state of the potential

$$V_0(x) = \frac{\hbar^2 x^2}{2mb_0^4} .$$

Suddenly the potential changes to

$$V_1(x) = \frac{\hbar^2}{2mb_1^4} (x-x_1)^2.$$

One wishes to find the probability the particle will be in any (say the ground) state of this new potential. What is involved in this case is a possible displacement to (x_1) and a change of frequency (from ω_0 to ω_1) of the potential, as illustrated in Figure 2.5.

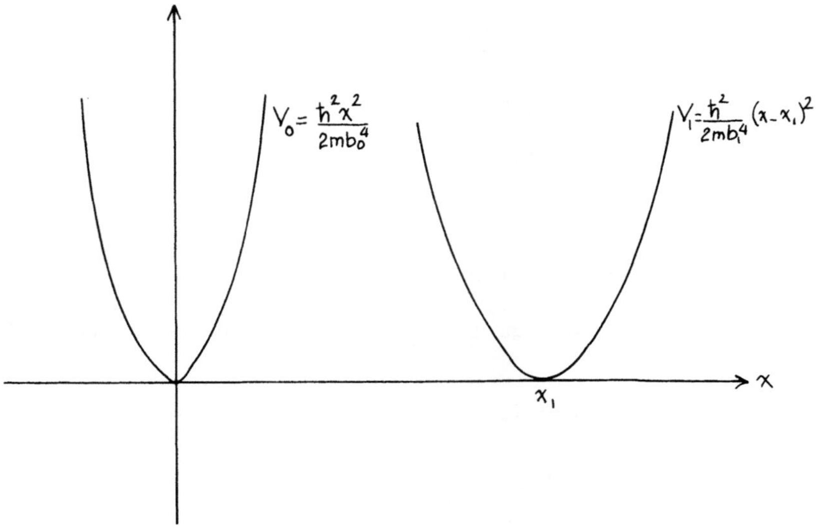

Figure 2.5. Change in potential in Example 2.8.

The wavefunction of the system is

$$\phi_0(x, b_0) = \left(\frac{1}{b_0\sqrt{\pi}}\right)^{\frac{1}{2}} e^{-x^2/2b_0^2}.$$

The potential changes frequency

$$\left(\omega = \frac{\hbar}{mb^2}\right)$$

and is displaced so the complete set of states which now describes the system is $\phi_n(x-x_1, b_1)$. To find the probability one merely uses Equation (2.3).

$$\phi_0(x,\ b_0) = \sum_{n=0}^{\infty} a_n\ \phi_n(x-x_1,\ b_1)$$

where

$$a_0 = \int_{-\infty}^{+\infty} \phi_0(x-x_1,\ b_1)\ \phi_0(x,b_0)dx = \frac{1}{\sqrt{b_0 b_1 \pi}} \int_{-\infty}^{+\infty} e^{\frac{-x^2}{2b_0^2} - \frac{(x-x_1)^2}{2b_1^2}} dx$$

$$= \sqrt{\frac{2b_0 b_1}{b_0^2+b_1^2}}\ e^{-\frac{x_1^2}{2(b_0^2+b_1^2)}} \quad ;$$

$$|a_0|^2 = \frac{2b_0 b_1}{b_0^2+b_1^2}\ e^{-\frac{x_1^2}{(b_0^2+b_1^2)}} \qquad (2.34)$$

Thus if $x_i = 0$ $|a_0|^2 = 0.8$ if $b_1 = 0.5b_0$ or $2b_0$ etc.

This procedure applies equally well if one changes the form of the potential.

EXAMPLE 2.9. Consider a particle wavefunction

$$\psi(x) = \frac{1}{4}\sqrt{\frac{15}{a^5}}\left(a^2-x^2\right); \quad |x| < a$$

$$= 0 \qquad |x| > a.$$

$$\left[\begin{array}{ll} \text{This involves the potential } V(x) = \dfrac{\hbar^2}{4ma^2}\dfrac{5x^2-a^2}{x^2-a^2} & |x| < a \\ \text{(c.f. Example 8.1.)} \qquad = \infty & |x| > a \end{array}\right].$$

The potential is suddenly changed to $V(x) = 0$ $|x| < a$, $V(x) = \infty$ $|x| > a$. What is the probability the particle is in the ground state of this potential? The normalized ground state of this new potential is just

$$\sqrt{\frac{1}{a}}\cos kx = \sqrt{\frac{1}{a}}\cos\frac{\pi x}{2a},$$

where

$$ka = \frac{\pi}{2}\ \text{(cf Eq. 2.8)}, \quad E = \frac{\hbar^2 k^2}{2m} = \frac{\hbar^2\pi^2/4}{2ma^2}.$$

Hence

$$a_0 = \frac{1}{4}\sqrt{\frac{15}{a^5}}\,\frac{1}{\sqrt{a}}\int_{-a}^{a}\cos\frac{\pi x}{2a}\left(a^2-x^2\right)dx = \frac{\sqrt{15}}{4a^3}\left[\frac{4a^3}{\pi} - \frac{8a^3}{\pi}\left(\frac{\pi^2}{2}-4\right)\right]$$

i.e.

$$a_0 = \frac{8\sqrt{15}}{\pi^3} = 0.9993 \text{ thus } |a_0|^2 = 0.9986 \simeq 1$$

as it should be since the wavefunction forms are very similar (cf. Example 8.1, Figure 8.1), as too their energies

$$\frac{\hbar^2\pi^2/4}{2ma^2} \quad \text{vs} \quad \frac{\hbar^2 5/2}{2ma^2} .$$

EXAMPLE 2.10. Consider a particle bound by the potential $-|V_0|\delta(x)$. The potential suddenly changes to $V = 0 \quad |x| < a$, $V = \infty |x| > a$. Find the probability the particle will be in one of the even parity states of this new potential.

Here

$$\phi_0 = \sqrt{\frac{|V_0|m}{\hbar^2}}\; e^{-\frac{|V_0|m|x|}{\hbar^2}}$$

and one wishes to evaluate the overlap

$$a_n = \int_{-a}^{a}\sqrt{\frac{|V_0|m}{\hbar^2}}\; e^{-\frac{|V_0|m}{\hbar^2}|x|}\sqrt{\frac{1}{a}}\cos\frac{(2n+1)\pi x}{2a}\,dx$$

$$= 2\left(\frac{|V_0|m}{\hbar^2 a}\right)^{\frac{1}{2}}\int_0^a e^{\frac{-|V_0|mx}{\hbar^2}}\cos\frac{\pi x}{2a}\,dx.$$

Evaluating this integral one obtains:

$$a_n = 2\left(\frac{|V_0|m}{\hbar^2 a}\right)^{\frac{1}{2}}\frac{\left[\frac{|V_0|m}{\hbar^2} + \frac{(2n+1)\pi}{2a}\,e^{-\frac{|V_0|ma}{\hbar^2}}\right]}{\frac{|V_0|^2m^2}{\hbar^4} + \left(\frac{(2n+1)\pi}{2a}\right)^2} .$$

As $a \to \infty$ $a_n \to 0$.

There is no probability the particle will be in one of the odd parity states of this potential, namely:

$$\phi_m = \sqrt{\frac{1}{a}} \sin \frac{m\pi x}{a}$$

EXAMPLE 2.11. A particle is originally in the ground state of the well

$$V = \infty \quad x < 0, \quad x > a,$$
$$= 0 \quad 0 < x < a.$$

Suddenly the wall at $x = a$ is shifted to $x = 2a$. Find the probability the particle will be in the ground state of the new potential which results from this shift.

The wavefunction is originally:

$$\phi_0 = \begin{cases} \sqrt{\frac{2}{a}} \sin \frac{\pi x}{a} & 0 < x < a, \\ 0 & x < 0, \quad x > a \end{cases}$$

and one is interested in the overlap:

$$a_0 = \int_0^a \sqrt{\frac{2}{a}} \sin \frac{\pi x}{a} \sqrt{\frac{1}{a}} \sin \frac{\pi x}{2a} \, dx$$

$$= 4\sqrt{2}/(3\pi).$$

Thus $|a_0|^2 = 32/(9\pi^2)$.

The original energy is $E = \hbar^2\pi^2/2ma^2$.

The energy of the new ground state is $\hbar^2\pi^2/8ma^2$ while that of the new first excited state is $\hbar^2\pi^2/2ma^2$, which is just the original energy. There is therefore a $32/9\pi^2$ probability the energy will be less than before.

The probability the energy will be unchanged is related to the overlap:

$$\int_0^a \sqrt{\frac{2}{a}} \sin \frac{\pi x}{a} \sqrt{\frac{1}{a}} \sin \frac{\pi x}{a} \, dx = \frac{1}{\sqrt{2}}.$$

Thus the probability the energy is unchanged is $\frac{1}{2}$. The probability the new energy is more than the original energy is

$$1 - \left(\frac{32}{9\pi^2} + \frac{1}{2}\right).$$

Expression (4.22) of Chapter 4 lists some other representations of the delta function.

REFERENCES

1. Higher Transcendental Functions, Erdélyi et al. McGraw Hill (1953) V. 2 p. 194.
2. Op cit. V. 1 p. 5.
3. Op cit. V. 2 p. 189, V. 3 p. 272.
4. Op cit. V. 2 p. 79.

CHAPTER 3

Momentum Space

Working in momentum space involves taking the Fourier transform of the eigenfunction $\psi(x, t)$ of the Schrödinger equation. Thus if

$$\phi(p, t) \equiv \frac{1}{\sqrt{2\pi}} \int_{-\infty}^{+\infty} e^{\frac{-ipx}{\hbar}} \psi(x, t)dx \tag{3.1}$$

it follows from the delta function property of Equation (2.19):

$$\frac{1}{2\pi} \int_{-\infty}^{+\infty} e^{\frac{iy(x-x')}{\hbar}} \frac{dy}{\hbar} = \delta(x-x'),$$

that

$$\psi(x, t) = \frac{1}{\sqrt{2\pi\hbar}} \int_{-\infty}^{+\infty} e^{\frac{ipx}{\hbar}} \phi(p, t)\, dp. \tag{3.2}$$

The function $\phi(p, t)$ is called the wavefunction "in momentum space". Assuming ψ and ϕ are normalizable (i.e. vanish at $\pm\infty$ so one can integrate by parts and drop surface terms), it can readily be shown that

$$p\phi = \frac{1}{\sqrt{2\pi}}\int e^{\frac{-ipx}{\hbar}} \frac{\hbar}{i}\frac{\partial\upsilon}{\partial x} dx; \quad -\frac{\hbar}{i}\frac{\partial}{\partial x}\phi = \frac{1}{\sqrt{2\pi}}\int e^{\frac{-ipx}{\hbar}} x\psi dx,$$

and for any operator $A(p)$

$$A(p)\phi(p, t) = \frac{1}{\sqrt{2\pi}} \int_{-\infty}^{+\infty} e^{\frac{-ipx}{\hbar}} A\left(\frac{\hbar}{i}\frac{\partial}{\partial x}\right) \psi(x, t)\, dx;$$

$$A\left(-\frac{\hbar}{i}\frac{\partial}{\partial p}\right)\phi(p,\ t) = \frac{1}{\sqrt{2\pi}}\int_{-\infty}^{+\infty} e^{-\frac{ipx}{\hbar}} A(x)\psi(x,\ t)\ dx \tag{3.3}$$

while similarly

$$A(x)\psi(x,\ t) = \frac{1}{\sqrt{2\pi}}\frac{1}{\hbar}\int_{-\infty}^{+\infty} e^{\frac{ipx}{\hbar}} A\left(-\frac{\hbar}{i}\frac{\partial}{\partial p}\right)\phi(p,\ t)\ dp;$$

$$B\left(\frac{\hbar}{i}\frac{\partial}{\partial x}\right)\psi(x,\ t) = \frac{1}{\sqrt{2\pi\hbar}}\int_{-\infty}^{+\infty} e^{\frac{ipx}{\hbar}} B(p)\phi(p,\ t)\ dp\ . \tag{3.4}$$

Also

$$\frac{\partial\phi(p,\ t)}{\partial t} = \frac{1}{\sqrt{2\pi}}\int e^{-\frac{ipx}{\hbar}} \frac{\partial}{\partial x}\psi(x,\ t)\ dx\ . \tag{3.5}$$

Thus

$$\left[\frac{p^2}{2m} + V\left(\frac{-\hbar}{i}\frac{\partial}{\partial p}\right) + \frac{\hbar}{i}\frac{\partial}{\partial t}\right]\phi(p,\ t) =$$

$$\frac{1}{\sqrt{2\pi}}\int e^{\frac{-ipx}{\hbar}}\left[-\frac{\hbar^2}{2m}\frac{\partial^2}{\partial x^2} + V(x) + \frac{\hbar}{i}\frac{\partial}{\partial t}\right]\psi(x,\ t)\ dx. \tag{3.6}$$

But since the integrand on the R.H.S. of Equation (3.6) vanishes, $\psi(x,\ t)$ being a solution of the Schrödinger equation, the momentum space function $\phi(p,\ t)$ satisfies an analogous equation:

$$\boxed{\left[\frac{p^2}{2m} + V\left(-\frac{\hbar}{i}\frac{\partial}{\partial p}\right) + \frac{\hbar}{i}\frac{\partial}{\partial t}\right]\phi(p,\ t) = 0,} \tag{3.7}$$

the Schrödinger equation "in momentum space".

Similarly expectation values can be equivalently evaluated in momentum space since

$$\int\psi^*(x,\ t)\Omega(x)\psi(x,\ t)\ dx = \frac{1}{\sqrt{2\pi\hbar}}\int_{x=-\infty}^{x=\infty}\int_{p'=-\infty}^{p'=\infty}\phi(p',\ t)\ e^{\frac{-ip'x}{\hbar}}$$

$$dp'\Omega(x)\psi(x,\ t)\ dx =$$

$$\frac{1}{2\pi\hbar^2}\int_{x=-\infty}^{x=\infty}\int_{p'=-\infty}^{p'=\infty}\phi(p', t)e^{\frac{-ip'x}{\hbar}}dp'\int_{p=-\infty}^{p=\infty}e^{\frac{ipx}{\hbar}}\Omega\left(-\frac{\hbar}{i}\frac{\partial}{\partial p}\right)\phi(p, t)\,dp\,dx =$$

$$\frac{1}{\hbar}\int_{p'=-\infty}^{p'=\infty}dp\,dp'\phi^*(p', t)\Omega\left(-\frac{\hbar}{i}\frac{\partial}{\partial p}\right)\phi(p, t)\,\frac{1}{2\pi}\int e^{i(p-p')\frac{x}{\hbar}}\frac{dx}{\hbar} = \int\phi^*(p, t)\Omega\left(\frac{-\hbar}{i}\frac{\partial}{\partial p}\right)\phi(p, t)\frac{dp}{\hbar} \quad (3.8)$$

One also notes that if $\psi(x, t)$ is normalized so is $\phi(p, t)$ since using Equation (3.8) with $\Omega = 1$:

$$\int_{-\infty}^{+\infty}\psi^*(x, t)\psi(x, t)\,dx = \int\phi^*(p, t)\phi(p, t)\,\frac{dp}{\hbar}\;.$$

The above formalism generalizes to three dimensions by replacing

$$\frac{\hbar}{i}\frac{\partial}{\partial x}\quad\text{by}\quad\frac{\hbar}{i}\nabla\ \text{etc.}$$

Six simple problems follow which illustrate the usefulness of the momentum representation and the fact that for certain potentials it is easier to work in momentum rather than coordinate space.

EXAMPLE 3.1. Consider the case of a free particle ($V(x) = 0$). In momentum space the Schrödinger equation, (Equation (3.7)) for this system is

$$\left\{\frac{p^2}{2m} - E\right\}\phi(p) = 0 \qquad (E > 0) \qquad (3.9)$$

i.e.

$$\phi(p) = \delta\left(\frac{p}{\hbar} - \frac{\sqrt{2mE}}{\hbar}\right)\;.$$

Since generally

$$\psi(x) = \frac{1}{\sqrt{2\pi}}\int_{-\infty}^{+\infty}\phi(p)e^{\frac{ipx}{\hbar}}\,d\left(\frac{p}{\hbar}\right)$$

if one suppresses the time variable in Equation (3.2), in this case

$$\psi(x) = \frac{1}{\sqrt{2\pi}}e^{i\frac{\sqrt{2mE}}{\hbar}x}\;,\qquad\text{as it should be.}$$

EXAMPLE 3.2. Consider a particle in the momentum dependent potential $V(x) = \alpha p$. In momentum space the Schrödinger equation for this potential (Equation (3.7)) is

$$\left\{\frac{p^2}{2m} + \alpha p - E\right\}\phi(p) = 0 \tag{3.10}$$

i.e.

$$\phi(p) = A\delta\left(\frac{p}{\hbar} - \frac{p_1}{\hbar}\right) + B\delta\left(\frac{p}{\hbar} - \frac{p_2}{\hbar}\right), \qquad \begin{array}{l} p_1 = -m\alpha + \sqrt{m^2\alpha^2 + 2mE} \\ p_2 = -m\alpha - \sqrt{m^2\alpha^2 + 2mE} \end{array}$$

and as above

$$\psi(x) = \frac{A}{\sqrt{2\pi}} e^{\frac{ip_1x}{\hbar}} + \frac{B}{\sqrt{2\pi}} e^{\frac{ip_2x}{\hbar}} .$$

EXAMPLE 3.3. Obtain the exact solution of the problem $V(x) = Ax$ $x > 0$ $(A > 0)$, $V(x) = \infty$ $x < 0$, (Drawn in Figure 3.1) with the help of the momentum representation.

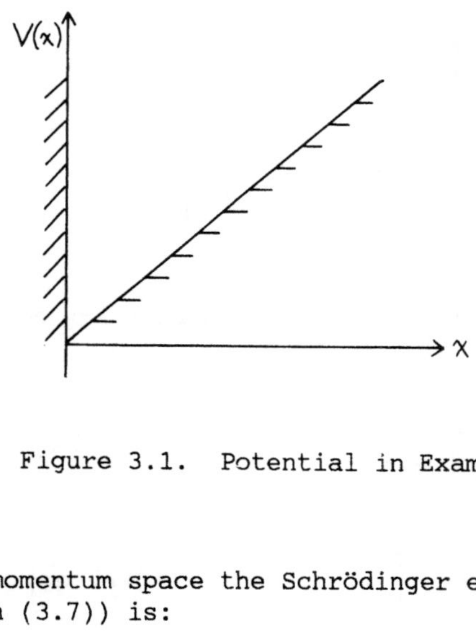

Figure 3.1. Potential in Example 3.3.

In momentum space the Schrödinger equation for this potential (Equation (3.7)) is:

$$\left\{\frac{p^2}{2m} - A\frac{\hbar}{i}\frac{d}{dp} - E\right\}\phi(p) = 0, \qquad (E > 0) \tag{3.11}$$

i.e.

$$\left\{\frac{p^2}{2m} - E\right\}\phi(p) = A\frac{\hbar}{i}\frac{d}{dp}\phi(p).$$

Integrating this expression yields:

$$\frac{i}{A\hbar}\left\{\frac{p^3}{6m} - Ep\right\} = \ln\left\{\phi(p)/C\right\},$$

i.e.

$$\phi(p) = Ce^{\frac{i}{A\hbar}\left\{\frac{p^3}{6m} - Ep\right\}} \qquad A,\ E > 0. \tag{3.12}$$

C can be obtained by normalization i.e. requiring

$$\int_{-\infty}^{+\infty} \phi_E^*(p)\phi_{E'}(p)\frac{dp}{\hbar} = \delta(E-E') = |C|^2 \int e^{i\left(\frac{E}{A\hbar} - \frac{E'}{A\hbar}\right)p} \frac{dp}{\hbar}$$

$\therefore |C|^2 2\pi A = 1$, where one has used Equation (2.19).
Thus within a phase

$$\phi(p) = \frac{1}{\sqrt{2\pi A}} e^{\frac{i}{A\hbar}\left\{\frac{p^3}{6m} - Ep\right\}}. \tag{3.13}$$

Generally

$$\psi(x) = \frac{1}{\sqrt{2\pi}} \int_{-\infty}^{+\infty} \phi(p)e^{\frac{ipx}{\hbar}} d(\tfrac{p}{\hbar}) \tag{3.14}$$

and in this case

$$\psi(x) = \frac{1}{2\pi\sqrt{A}} \int_{p=-\infty}^{p=\infty} e^{\frac{i}{A\hbar}\left\{\frac{p^3}{6m} - Ep\right\} + \frac{ipx}{\hbar}} d\left(\frac{p}{\hbar}\right). \tag{3.15}$$

For this problem the wavefunction $\psi(x)$ satisfies the boundary condition $\psi(0) = 0$, since $V(0) = \infty$. Hence for this example

$$\psi(0) = \frac{1}{\hbar\sqrt{2\pi}} \int_{-\infty}^{+\infty} \phi(p)\ dp = \frac{1}{2\pi\sqrt{A}} \int_{-\infty}^{+\infty} e^{\frac{i}{A\hbar}\left\{\frac{p^3}{6m} - Ep\right\}} d\left(\frac{p}{\hbar}\right) = 0.$$

$$= \frac{1}{2\pi\sqrt{A}\hbar}\left[\int_{-\infty}^{+\infty} \cos(p^3/6m - Ep)/A\hbar\ dp + i\int_{-\infty}^{+\infty} \sin(p^3/6m - Ep)/A\hbar\ dp\right].$$

This implies the even integral,

$$\int_0^\infty \cos(p^3/6m - Ep)/A\hbar \, dp = 0 \tag{3.16}$$

since the integrand of the integral

$$\int_{-\infty}^{+\infty} \sin(p^3/6m - Ep)/A\hbar \, dp$$

is odd making the latter integral automatically zero.
Defining $u \equiv p/(2mA\hbar)^{1/3}$ the integral in Equation (3.16) becomes

$$\int_0^\infty \cos(u^3/3 - E(2m/A^2\hbar^2)^{1/3}u) \, du = \sqrt{\pi}\, \Phi\left(-E\left\{2m/A^2\hbar^2\right\}^{1/3}\right),$$

where $\Phi\left(-E\left\{2m/A^2\hbar^2\right\}^{1/3}\right)$ is the Airy function.

The energy eigenvalues E in this problem thus satisfy the condition that

$$\Phi\left(-E_n\left\{2m/A^2\hbar^2\right\}^{1/3}\right) = 0. \tag{3.17}$$

For reasonably large negative arguments

$$\Phi(x) \to \frac{1}{|x|^{1/4}} \sin\left(\frac{2}{3}|x|^{3/2} + \frac{\pi}{4}\right).$$

Hence in this limit

$$\frac{2}{3} E_n^{3/2} \left\{\frac{2m}{A^2\hbar^2}\right\}^{1/2} + \frac{\pi}{4} = n\pi \qquad n = 1, 2\ldots$$

i.e.

$$E_n = (n-\tfrac{1}{4})^{2/3} \left(\frac{A^2\hbar^2}{m}\right)^{1/3} \left(\frac{3\pi}{2\sqrt{2}}\right)^{2/3} \qquad n = 1, 2\ldots \tag{3.18}$$

which can be compared with

$$E_n = n^{2/3}\left(\frac{A^2\hbar^2}{m}\right)^{1/3} \left(\frac{3\pi}{2\sqrt{2}}\right)^{2/3},$$

the Wilson-Sommerfeld result (Equation (1.12)).

EXAMPLE 3.4. Obtain the exact solution for a particle in the potential

$$V(x) = A/x \qquad x > 0 \qquad (A < 0),$$

$$V(x) = \infty \qquad x < 0,$$

using the momentum representation. This potential is illustrated in Figure 3.2.

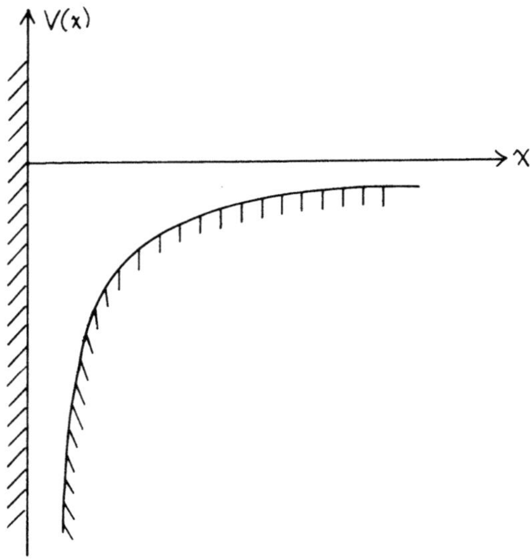

Figure 3.2. Potential in Example 3.4.

In momentum space the Schrödinger equation for this potential is:

$$\left\{\frac{p^2}{2m} + \frac{A}{\frac{\hbar}{i}\frac{d}{dp}} - E\right\} \phi(p) = 0, \qquad (E < 0)\ldots \tag{3.19}$$

i.e.

$$-\frac{\left(\frac{iA}{\hbar} - \frac{p}{m}\right)}{E - \frac{p^2}{2m}}\,dp = \frac{d\phi(p)}{\phi(p)} = -\frac{d\left(\frac{p^2}{2m} + |E|\right)}{\frac{p^2}{2m} + |E|} + \frac{\frac{iA}{\hbar}\,dp}{\frac{p^2}{2m} + |E|}\ .$$

Integrating this expression yields:

$$\ell n\left\{\phi(p)(|E| + p^2/2m)/C\right\} = -\frac{i|A|}{\hbar}\sqrt{\frac{2m}{|E|}}\tan^{-1}\frac{p}{\sqrt{2m|E|}}$$

i.e.

$$\phi(p) = \frac{C}{|E| + p^2/2m}\, e^{-\frac{i|A|}{\hbar}\sqrt{\frac{2m}{|E|}}\tan^{-1}\frac{p}{\sqrt{2m|E|}}}. \tag{3.20}$$

$|\phi(p)|^2$ vanishes at $\pm\infty$ and is square normalizable. Thus C can be determined by normalization i.e.

$$\int_{-\infty}^{+\infty} |\phi(p)|^2 \, d\frac{p}{\hbar} = 1 = |C|^2 \int \frac{dp/\hbar}{(|E| + p^2/2m)^2}$$

$$\therefore \frac{|C|^2}{\hbar}\sqrt{\frac{\pi^2 m}{2|E|^3}} = 1.$$

Thus to within a phase:

$$\phi(p) = \frac{\left(\frac{2\hbar^2|E|^3}{\pi^2 m}\right)^{1/4}}{|E| + p^2/2m}\, e^{-\frac{i|A|}{\hbar}\sqrt{\frac{2m}{|E|}}\tan^{-1}\frac{p}{\sqrt{2m|E|}}}. \tag{3.21}$$

Generally

$$\psi(x) = \frac{1}{\sqrt{2\pi}}\int_{-\infty}^{+\infty} \phi(p)e^{\frac{ipx}{\hbar}} \, d\left(\frac{p}{\hbar}\right).$$

In this case $\psi(0) = 0$, since $V(0) = \infty$.

Hence

$$0 = \frac{1}{\pi}\left(\frac{|E|^3}{2m\hbar^2}\right)^{1/4}\int_{-\infty}^{+\infty} \frac{1}{|E|+p^2/2m}\left(\cos\frac{|A|}{\hbar}\sqrt{\frac{2m}{|E|}}\tan^{-1}\frac{p}{\sqrt{2m|E|}}\right.$$

$$\left. - i\sin\frac{|A|}{\hbar}\sqrt{\frac{2m}{|E|}}\tan^{-1}\frac{p}{\sqrt{2m|E|}}\right) dp.$$

The integral

$$\int_{-\infty}^{+\infty} \frac{1}{|E|+p^2/2m} \sin\left(|A|\sqrt{\frac{2m}{|E|}} \tan^{-1} \frac{p}{\sqrt{2m|E|}}\right) dp$$

is automatically zero since the integrand is odd. Thus the even integral

$$\int \frac{\cos\left(\frac{|A|}{\hbar}\sqrt{\frac{2m}{|E|}} \tan^{-1} \frac{p}{\sqrt{2m|E|}}\right)}{|E| + p^2/2m} dp$$

must be zero in order for $\psi(0)$ to be zero.
Integrating this expression one obtains

$$\frac{\hbar}{|A|} \sin \sqrt{\frac{2m}{|E|}} \frac{|A|}{\hbar} \theta \Big|_0^{\pi/2} = \frac{\hbar}{|A|} \sin \sqrt{\frac{2m}{|E|}} \frac{|A|\pi}{\hbar 2} . \tag{3.22}$$

Hence

$$\sin \frac{|A|\pi}{2\hbar} \sqrt{\frac{2m}{|E|}} = 0.$$

The energy eigenvalues $|E_n|$ of this problem thus satisfy the equation

$$\frac{\pi|A|}{2\hbar} \sqrt{\frac{2m}{|E_n|}} = n\pi \qquad n = 1, 2...$$

or

$$E_n = -\frac{|A|^2 m}{2\hbar^2 n^2} . \tag{3.23}$$

With $|A| = \alpha\hbar c$ this reduces to the usual Bohr result for the states u_{no} (r) (as well as for the states u_{no} (r) since the energies in the hydrogen atom are independent of ℓ, an effect known as an "accidental" degeneracy).

Substituting expression (3.23) into expression (3.21) one obtains a general expression for the wavefunction in the momentum representation:

$$\phi_n(p) = \frac{\left(\frac{2\hbar}{\pi}\beta_n\right)^{1/2} e^{-2in \tan^{-1}\beta_n p}}{1 + (\beta_n p)^2} , \quad \beta_n \equiv \frac{n\hbar}{m|A|} \tag{3.24}$$

$$n = 1, 2 \ldots .$$

Using this expression one readily obtains:

$$<\phi_n(p)p^{2m+1}\,\phi_n(p)> = 0 \qquad m = 0,\ 1\ldots\ ,$$

$$<p^2> = <\phi_n(p)p^2\phi_n(p)> = \frac{m^2|A|^2}{n^2\hbar^2}\ , \tag{3.25}$$

$$<\frac{p^2}{2m}> = -\ E_n .$$

These two equations imply

$$\Delta p = \sqrt{<p^2> - <p>^2} = \frac{m|A|}{n\hbar}\ . \tag{3.26}$$

Similarly one can easily obtain:

$$<r> = <\phi_n(p)r\phi_n(p)> = \frac{3\hbar^2n^2}{2m|A|}\ \ldots\ , \tag{3.27}$$

$$<r^2> = \frac{\hbar^4}{2}\,(5n^2+1)\,\frac{n^2}{m^2|A|^2}\ \ldots\ . \tag{3.28}$$

These results imply:

$$\Delta r = \sqrt{<r^2> - <r>^2} = \frac{n\hbar^2}{2m|A|}\,(n^2+2)^{1/2}. \tag{3.29}$$

Combining Equations (3.26) and (3.29) one obtains:

$$\Delta r\Delta p = \frac{\hbar}{2}\sqrt{n^2+2} > \frac{\hbar}{2}\ \ldots \tag{3.30}$$

consistent with the uncertainty relation (cf. Chapter 4).

EXAMPLE 3.5. Solve exactly the problem of a particle in the potential $V(x) = A|x|\quad -\infty < x < \infty$.

EXAMPLE 3.6. Solve exactly the problem of a particle in the potential $V(x) = \frac{A}{|x|}\quad -\infty < x < \infty$.

The potentials of Examples 3.5 and 3.6 are symmetric potentials i.e. if one plots V(x) vs x the potential for $x \le 0$ is the mirror reflection (about the V(x) axis) of the wells for $x > 0$ in Examples 3.3 and 3.4 (Figures 3.1 and 3.2). One can proceed here using methods similar to those used in Example 3.3 and 3.4. However, instead if one compares (cf. Equations (1.8) and (1.10) the solutions of the problem of the standard harmonic oscillator potential restricted to $x > 0$ i.e. $V(x) = Ax^2$ $0 < x < \infty$, $= \infty$ $x \le 0$, to the oscillator extending over all space $V(x) = Ax^2$, $-\infty < x < \infty$, the former potential's allowed energies are

$$E_n = (2n - \tfrac{1}{2})\hbar\sqrt{\frac{2A}{m}} \qquad n = 1, 2 \ldots$$

and correspond to odd parity solutions (see Chapter 10) while the latter's allowed energies are

$$E_n = (n - \tfrac{1}{2})\hbar\sqrt{\frac{2A}{m}} \qquad n = 1, 2 \ldots$$

and both odd and even parity solutions are allowed. One can thus obtain the solutions in the latter case from the solutions in the former by letting $n \to n/2$. Similarly if one considers the infinite well $V(x) = \infty$ $x < 0$, $x > a$, $V(x) = 0$ $0 < x < a$ its solutions are the odd parity wavefunctions $\psi(x) = \sqrt{2/a}\sin kx$ (where $ka = n\pi$) i.e.

$$E = \frac{\hbar^2k^2}{2m} = \frac{\hbar^2n^2\pi^2}{2ma} \qquad n = 1, 2 \ldots$$

whereas if one considers the infinite well $V(x) = \infty$ $x < -a$, $x < a$, $V(x) = 0$, $-a < x < a$, its solutions are $\psi(x) = \sqrt{1/a}\sin kx$ (odd parity) and $\psi(x) = \sqrt{1/a}\cos kx$ (even parity) where $ka = n\pi$ or $(n-\tfrac{1}{2})\pi$ $n = 1, 2 \ldots$ i.e.

$$E = \frac{\hbar^2}{2m}\left(\frac{n}{2}\right)^2\frac{\pi^2}{a^2} \qquad n = 1, 2 \ldots$$

Again the solutions for the latter case can be obtained from the former by letting $n \to n/2$. A third example which illustrates the fact that this procedure may be applied generally is the well $V(x) = -V_0\cosh^{-2}ax^{1})$ $x > 0$, $V(x) = \infty$ $x < 0$. The (odd parity) solutions for this case are:

$$\psi(x) = (1-\xi^2)^{(s-2n+1)/2}\ {}_2F_1(1-2n,\ 2s-2n+2;\ s-2n+2,\ (1-\xi)/2)$$

with corresponding energies

$$E_n = \frac{\hbar^2a^2}{8m}\left[-(4n-1) + \sqrt{1 + \frac{8mV_0}{a^2\hbar^2}}\right]^2, \quad n = 1, 2 \ldots$$

while the symmetric potential $V(x) = -V_0\cosh^{-2}ax$ (all x) has solutions

$$\psi = (1-\xi^2)^{(s-n+1)/2}\ {}_2F_1(1-n,\ 2s-n+2;\ s-n+2;\ (1-\xi)/2)$$

where

$$E_n = -\frac{\hbar^2a^2}{8m}\left[-(2n-1) + \sqrt{1 + \frac{8mV_0}{a^2\hbar^2}}\right]^2, \qquad n = 1, 2 \ldots$$

(with the symbols s, and ξ defined as

$$s = \tfrac{1}{2}\left(-1 + \sqrt{1 + \frac{8mV_0}{a^2\hbar^2}}\right), \quad \xi = \tanh ax).$$

Thus in this case also one obtains the solutions for the symmetric well by letting $n \to n/2$ in the solutions for the well which extends only from $0 < x < \infty$.

The solutions for Example 3.3 are

$$E_n = (n-\tfrac{1}{4})^{2/3}\left(\frac{A^2\hbar^2}{m}\right)^{1/3}\left(\frac{3\pi}{2\sqrt{2}}\right)^{2/3}$$

$$= (2n-\tfrac{1}{2})^{2/3}\left(\frac{A^2\hbar^2}{m}\right)^{1/3}\left(\frac{3\pi}{4\sqrt{2}}\right)^{2/3} \qquad n = 1, 2 \ldots .$$

These correspond only to the odd parity solutions of Example 3.5. By analogy with the above three problems one thus expects that the odd and even parity solutions for Example 3.5 have energies:

$$E_n = (n-\tfrac{1}{2})^{2/3}\left(\frac{A^2\hbar^2}{m}\right)^{1/3}\left(\frac{3\pi}{4\sqrt{2}}\right)^{2/3} \qquad n = 1, 2 \ldots . \tag{3.31}$$

which for large n agrees with the Wilson-Sommerfeld result (Equation (1.11)).

Similarly the solution for Example 3.4 is

$$E_n = -\frac{|A|^2 m}{2n^2\hbar^2} = -\frac{2|A|^2 m}{(2n)^2\hbar^2}$$

and one expects that the energy in Example 3.6 is just

$$E_n = -\frac{2|A|^2 m}{n^2\hbar^2} \tag{3.32}$$

and that this includes both even and odd parity solutions. This result agrees precisely with the Wilson-Sommerfeld result (Equation (1.13)).

REFERENCE

1. L. Landau et I. Lifchitz, Mechanique Quantique (Eds Mir) Moscow 1966, p. 94.

CHAPTER 4

Wavepackets and the uncertainty principle

The properly normalized free particle wavefunction is

$$\psi_p(x, t) = \frac{1}{\sqrt{2\pi}} e^{i\left(\frac{px}{\hbar} - \frac{Et}{\hbar}\right)} \quad \text{(cf. Equation (2.18)} \tag{4.1}$$

One problem with this function is that it has no spatial localization i.e. though the momentum is precisely known (in other words $\Delta p = 0$), $\Delta x = \infty$, where ΔA implies uncertainty in A.

The shortcoming may be easily removed by constructing a wavepacket

$$\Psi(x, t) = \int_{p'=-\infty}^{p'=\infty} A(p')\, \psi_{p'}(x, t)\, \frac{dp'}{\hbar}$$

$$= \frac{1}{\sqrt{2\pi}} \int_{p'=-\infty}^{p'=\infty} A(p')\, e^{\frac{i}{\hbar}(p'x - E(p')t)}\, \frac{dp'}{\hbar} \tag{4.2}$$

where $A(p')$ is a function concentrated about $p' = p$. Thus if

$$A(p') = \delta\left(\frac{p'-p}{\hbar}\right)$$

this reduces to expression (4.1). The function

$$A(p')\, e^{-\frac{iE(p')t}{\hbar}}$$

is in fact the wavefunction in momentum space (cf. Equation (3.2)).

Any detailed functional form $A(p')$ gives the same general properties to $\Psi(x, t)$ namely localizes the particle, $(\Delta x \neq \infty)$.

EXAMPLE 4.1. Suppose

$$A(p') = \sqrt{\frac{\hbar}{\delta}}\ e^{-|p'-p|/\delta}\ , \tag{4.3}$$

a normalized function peaked about $p' = p$ and illustrated in Figure 4.1.

As first sight it appears this gives an uncertainty in momentum $\Delta p \sim 2\delta$.

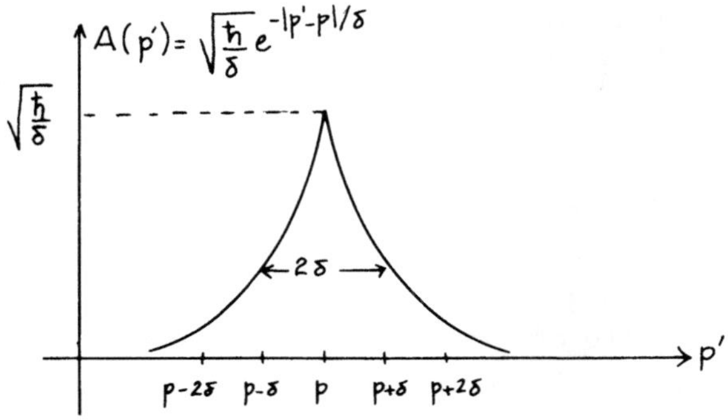

Figure 4.1. The momentum distribution in Example 4.1.

The gain however is that $\Psi(x, t)$ is now localized. Thus

$$\Psi(x, 0) = \sqrt{\frac{\hbar}{2\pi\delta}} \int_{p'=-\infty}^{p'=\infty} e^{-|p'-p|/\delta + \frac{ip'x}{\hbar}}\ \frac{dp'}{\hbar}\ ,$$

or

$$\Psi(x, 0) = \sqrt{\frac{2\hbar^3}{\delta^3\pi}}\ e^{\frac{ipx}{\hbar}}\ \frac{1}{x^2 + \frac{\hbar^2}{\delta^2}}\ , \tag{4.4}$$

[The normalization of $\Psi(x, 0)$ can be confirmed by noting

$$\int_{-\infty}^{+\infty} \frac{dx}{(x^2+1)^2} = \frac{\pi}{2}\ ]$$

The amplitude of the function of Equation (4.4) (ignoring the phase) is drawn in Figure 4.2. There is a spread about $x = 0$ given approximately

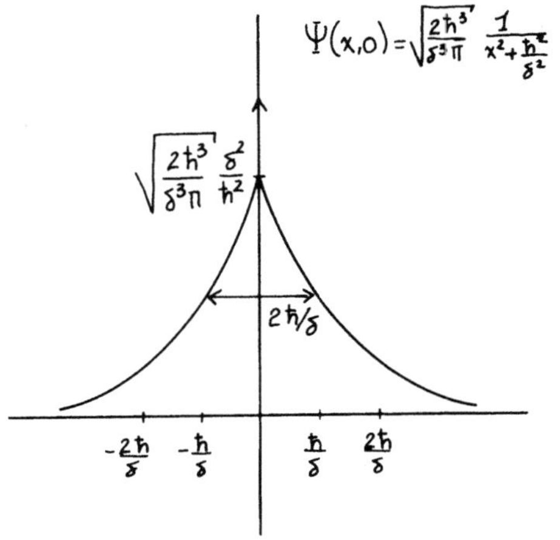

Figure 4.2. Amplitude of $\Psi(x, 0)$ in Example 4.1.

by $\Delta x = 2\hbar/\delta$. Thus $\Delta x\Delta p \sim 2\hbar/\delta \; 2\delta = 4\hbar$.

A more accurate estimate of $\Delta x\Delta p$ is possible since $(\Delta O)^2 = \langle O^2\rangle - \langle O\rangle^2$ where $\langle O\rangle = \int \Psi^* O\Psi d\tau / \int \Psi^*\Psi d\tau = \int A^*(p) O A(p) dp / \int A^*(p) A(p) dp$.

Using the function of Equation (4.4), $\langle x\rangle = 0$, while

$$(\Delta x)^2 = \langle x^2\rangle = \frac{2\hbar^3}{\delta^3\pi}\int_{x=-\infty}^{x=\infty} \frac{x^2 dx}{\left(x^2 + \frac{\hbar^2}{\delta^2}\right)^2} = \frac{\hbar^2}{\delta^2} .$$

Similarly

$$\langle p\rangle = \frac{\hbar}{\delta}\int_{p'=-\infty}^{p'=\infty} p' \, e^{-|p'-p|/\delta} \frac{dp'}{\hbar} = p,$$

while

$$\langle p^2\rangle = p^2 + \frac{\delta^2}{2} \quad \text{i.e. } \Delta p = \delta/\sqrt{2} ,$$

and

$$\Delta x\Delta p = \frac{\hbar}{\delta}\frac{\delta}{\sqrt{2}} = \frac{\hbar}{\sqrt{2}} \simeq 0.71 \, \hbar, \qquad (4.5)$$

in this case. Thus the uncertainty in the position multiplied by the uncertainty in the momentum of a localized wavepacket is of the order of $\hbar$. Other A(p)'s yield similar results.

EXAMPLE 4.2. Consider the case

$$A(p') = \sqrt{\frac{3\hbar}{2\delta^3}}\left[\delta - |p'-p|\right], \tag{4.6}$$

$$|p'-p| < \delta, \qquad A(p') = 0 \qquad |p'-p| > \delta,$$

a normalized function peaked at p' = p. This A(p') is illustrated in Figure 4.3.

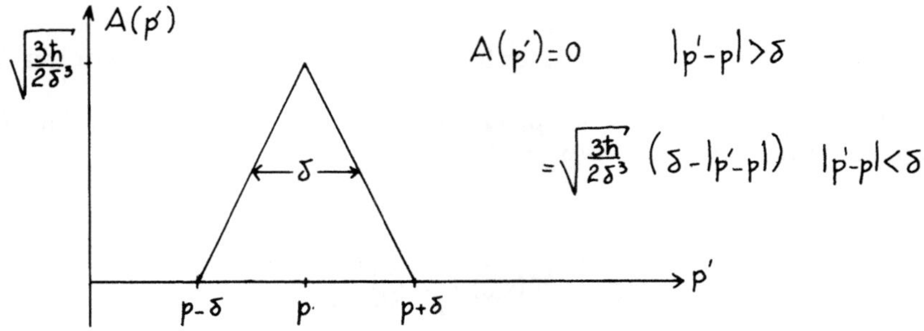

Figure 4.3. The momentum distribution in Example 4.2.

A first sight it appears this gives an uncertainty $\Delta p \sim \delta$. A more accurate estimate of Δp can be carried out by evaluating

$$\langle p\rangle = \frac{3\hbar}{2\delta^3}\int_{p-\delta}^{p+\delta} (\delta-|p'-p|)^2 p' \,\frac{dp'}{\hbar} = p$$

$$\langle p^2\rangle = \frac{3\hbar}{2\delta^3}\int_{p-\delta}^{p+\delta} (\delta-|p'-p|)^2\, p'^2 \,\frac{dp'}{\hbar} = p^2 + \frac{\delta^2}{10}\,.$$

Thus

$$\Delta p = \frac{\delta}{\sqrt{10}}\ .$$

In this case

$$\Psi(x,\,0) = \frac{1}{\sqrt{2\pi}}\sqrt{\frac{3\hbar}{2\delta^3}}\int_{p-\delta}^{p+\delta} (\delta-|p'-p|)\, e^{ip'x/\hbar}\,\frac{dp'}{\hbar}\ .$$

Evaluating this integral one obtains:

$$\Psi(x,\ 0) = \frac{1}{2}\sqrt{\frac{3\delta}{\pi\hbar}}\ e^{\frac{ipx}{\hbar}}\ \frac{\sin^2 \frac{\delta x}{2\hbar}}{\left(\frac{\delta x}{2\hbar}\right)^2} . \tag{4.7}$$

[The normalization of this expression may easily be verified using the result

$$\int_0^\infty du\ \frac{\sin^4 u}{u^4} = \frac{\pi}{3} .$$
]

The amplitude of the function of Equation (4.7) (ignoring the phase) is drawn in Figure 4.4.

There appears to be a spread about x of approximately $2\hbar\pi/\delta$ giving an approximate $\Delta x \Delta p \sim 2\hbar\pi/\delta \quad \delta = h$.

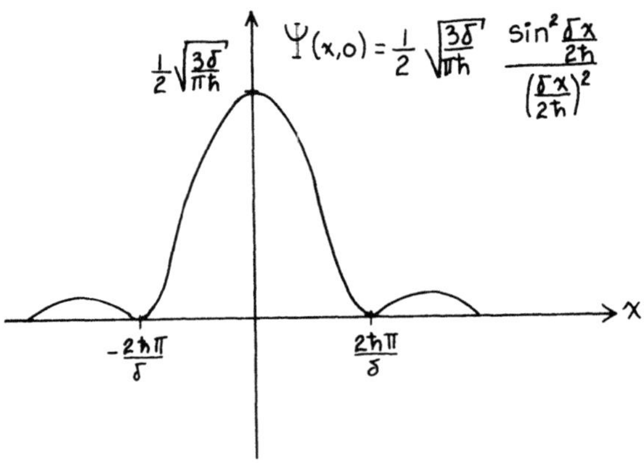

Figure 4.4. Amplitude of the wavefunction $\Psi(x,\ 0)$ in Example 4.2.

For a more accurate Δx one evaluates

$$\langle x\rangle = \frac{1}{4}\ \frac{3}{\pi}\ \frac{\delta}{\hbar} \int_{-\infty}^{+\infty} \frac{\sin^4 \frac{\delta x}{2\hbar}}{\left(\frac{\delta x}{2\hbar}\right)^4}\ x\ dx = 0,$$

$$\langle x^2\rangle = \frac{1}{4}\ \frac{3}{\pi}\ \frac{\delta}{\hbar} \int_{x=-\infty}^{x=\infty} \frac{\sin^4 \frac{\delta x}{2\hbar}}{\left(\frac{\delta x}{2\hbar}\right)^4}\ x^2 dx = \frac{6\hbar^2}{\pi\delta^2} \int_{-\infty}^{+\infty} \frac{\sin^4 u}{u^2}\ du$$

$$= \frac{6\hbar^2}{\pi\delta^2}\left[\int_{-\infty}^{+\infty}\frac{\sin^2 u}{u^2}\,du - \frac{1}{4}\int_{-\infty}^{\infty}\frac{\sin^2 2u}{u^2}\,du\right] = \frac{3\hbar^2}{\delta^2}\,.$$

Hence,

$$\Delta x = \frac{\sqrt{3}\hbar}{\delta}\,, \quad \text{and} \quad \Delta x\Delta p = \frac{\sqrt{3}\,\hbar}{\delta}\,\frac{\delta}{\sqrt{10}} = \sqrt{\frac{3}{10}}\,\hbar \approx 0.55\hbar$$

in this case, a somewhat smaller result than that obtained in Example 4.1.

[The integral $\int_0^{\infty}\frac{\sin^2 u\,du}{u^2} = \frac{\pi}{2}$ is needed to obtain the above result.]

Obviously $\psi(x, t)$ of Equation (4.2) also satisfies the free particle Schrödinger equation just as Equation (4.1) does namely:

$$-\frac{\hbar^2}{2m}\frac{\partial^2}{\partial x^2}\Psi(x, t) = -\frac{\hbar}{i}\frac{\partial}{\partial t}\Psi(x, t), \tag{4.8}$$

independent of the detailed form of A(p') since

$$\frac{1}{\sqrt{2\pi}}\int_{-\infty}^{+\infty} A(p')\left(\frac{p'^2}{2m} - E(p')\right) e^{\frac{i}{\hbar}(p'x - E(p')t)}\,\frac{dp'}{\hbar} = 0,$$

if

$$\frac{p^2}{2m} = E(p).$$

Further, independent of the form of A(p), since it is the wave-function in momentum space, one can write:

$$A(p') = \frac{1}{\sqrt{2\pi}}\int \Psi(x', 0)\, e^{-\frac{ip'x'}{\hbar}}\,dx'. \tag{4.9}$$

Substituting Equation (4.9) into Equation (4.2), and dropping the prime for the p's one obtains:

$$\Psi(x, t) = \int\Psi(x', 0)\frac{1}{2\pi}\int e^{i\left[\frac{(x-x')p}{\hbar} - \frac{E(p)t}{\hbar}\right]}\,\frac{dp}{\hbar}\,dx' \tag{4.10}$$

Defining

$$G(x,\ x',\ t)\ \equiv \frac{1}{2\pi}\int e^{i\left[\frac{(x-x')}{\hbar}p\ -\ \frac{E(p)t}{\hbar}\right]}\ \frac{dp}{\hbar} \tag{4.11a}$$

$$= \int \phi_p^*(x')\phi_p(x)\ e^{-\frac{iE(p)t}{\hbar}}\ dp/\hbar \tag{4.11b}$$

where $\phi_p(x)$ are the properly normalized free particle wavefunctions (cf. Equation (2.18)), one can write Equation (4.10) as follows:

$$\Psi(x,\ t) = \int \Psi(x',\ 0)\ G(x,\ x',\ t)\ dx'. \tag{4.12}$$

The function G(x, x', t) of Equation (4.11) known as the free particle Green's function can be evaluated explicitly by integrating expression (4.11):

$$G(x,\ x',\ t) = \left(\frac{m}{2\pi i\hbar t}\right)^{1/2}\ \exp\left(-\ \frac{m(x-x')^2}{2i\hbar t}\right). \tag{4.13}$$

This function also satisfies the Schrödinger equation Equation (4.8) and for $t \to 0$

$$G(x,\ x',\ 0) = \delta(x-x'),$$

(Equation (4.13) having in this case the δ function form of Equation (2.15)) as it must since $\Psi(x,\ 0) = \int\Psi(x',\ 0)G(x,\ x',\ 0)\ dx'$, in this case.

A formula useful in calculating reflection and transmission times for wavepackets, which is a result independent of the details of A(p') (but assuming it is peaked about p' = p) involves expanding E(p') about p' = p in expression (2)

$$E(p') = \frac{p'^2}{2m} = \frac{p^2}{2m} + \left.\frac{dE(p')}{dp'}\right|_{p'=p}\ (p'-p) + \ \ldots$$

Keeping only linear terms (i.e. assuming only values about p' = p are important) yields:

$$\Psi(x,\ t) = \frac{1}{\sqrt{2\pi}}\int_{p'=-\infty}^{p'=\infty} A(p')\ e^{\frac{i}{\hbar}\left(p'x-\frac{p^2}{2m}t+\left.\frac{dE(p')}{dp'}\right|_{p'=p}tp\ -\ \left.\frac{dE(p')}{dp'}\right|_{p'=p}tp'\right)}\frac{dp'}{\hbar}$$

$$= \frac{1}{\sqrt{2\pi}}\ e^{-\frac{i}{\hbar}\left(\frac{p^2}{2m}\ t\ -\ \left.\frac{dE(p')}{dp'}\right|_{p'=p}tp\right)}\int_{p'=-\infty}^{p'=\infty} A(p')\ e^{\frac{i}{\hbar}\ p'\left(x\ -\ \left.\frac{dE(p')}{dp'}\right|_{p'=p}t\right)}\ \frac{dp'}{\hbar}\ .$$

Thus

$$\Psi(x, t) = e^{i\phi}\, \Psi\left(x - \left.\frac{dE(p')}{dp'}\right|_{p'=p} t, 0\right) \tag{4.14}$$

Equation (4.14), which is a good approximation if A(p') is peaked about p' = p, and is independent of the details of A(p'), implies the wavepacket $\Psi(x, t)$ moves with a speed

$$\left.\frac{dE(p')}{dp'}\right|_{p'=p} .$$

Since

$$E(p') = \frac{p'^2}{2m}, \quad \frac{dE(p')}{dp'} = \frac{p'}{m} = v_g .$$

This is known as the group velocity of the wavepacket and is just the classical speed of a free particle.

EXAMPLE 4.3. Given

$$A(p') = \sqrt{\frac{3\hbar}{2\pi\delta}}\; \frac{\sin^2 (p'-p)/\delta}{((p'-p)/\delta)^2} \tag{4.15}$$

find $\Psi(x, 0)$.

One can proceed as in the previous examples or note that since, according to Equation (4.2) (where we use k rather than p to make the equations look a little more symmetric)

$$\Psi(x, 0) = \frac{1}{\sqrt{2\pi}} \int A(k') e^{ik'x}\, dk', \qquad (\hbar k = p)$$

if

$$\Psi(x, 0) = e^{ikx}\, \phi(x)$$

$$\phi(x) = \frac{1}{\sqrt{2\pi}} \int A(k') e^{i(k'-k)x} dk' \tag{4.16}$$

where according to Equation (4.9)

$$A(k') = \frac{1}{\sqrt{2\pi}} \int \Psi(x', 0)\, e^{-ik'x'}\, dx' = \frac{1}{\sqrt{2\pi}} \int \phi(x')\, e^{-ix'(k'-k)} dx' .$$

Hence

$$A(k'+k) = \frac{1}{\sqrt{2\pi}} \int \phi(x')\, e^{-ix'k'} dx' = \frac{1}{\sqrt{2\pi}} \int \phi(x') e^{ix'k'}\, dx'$$

if $\phi(x)=\phi(-x)$.

or

$$A(k'+k) = \frac{1}{\sqrt{2\pi}} \int \phi(x'-x)\, e^{i(x'-x)k'}\, dx'. \qquad (4.17)$$

Comparing Equations (4.16) and (4.17) one notes that if one identifies $A(k')$ with a particular (even) function $\phi(x'-x)$ one can then identify $\phi(x)$ with the corresponding $A(k + k')$.
In Example 4.2

$$\phi(x) = \frac{1}{2}\sqrt{\frac{3\delta}{\pi\hbar}}\ \frac{\sin \delta x/2\hbar}{\left(\frac{\delta x}{2\hbar}\right)^2} \quad \text{and} \quad A(k+k') = \sqrt{\frac{3\hbar}{2\delta^3}}\,[\delta-|k'|\hbar].$$

Hence if one identifies $A(k')$ with

$$\frac{1}{2}\sqrt{\frac{3\delta}{\pi\hbar}}\ \sin^2 \frac{(k'-k)\delta/2\hbar}{((k'-k)\delta/2\hbar)^2}\ ,$$

one obtains the corresponding

$$\phi(x) = \sqrt{\frac{3\hbar}{2\delta^3}}\,[\delta - |x|\hbar]$$

directly from $A(k'+k)$.

Letting $\delta \to 2\hbar^2/\delta$

$$A(k') = \sqrt{\frac{3\hbar}{2\pi\delta}}\ \frac{\sin^2(p'-p)/\delta}{((p'-p)/\delta)^2}$$

and

$$\psi(x, 0) = e^{\frac{ipx}{\hbar}}\ \frac{1}{4}\sqrt{\frac{3\delta}{\hbar}}\left[2 - \frac{|x|\delta}{\hbar}\right]. \qquad (4.18)$$

EXAMPLE 4.4. Consider the case

$$A(p') = \sqrt{\frac{2\delta^3\hbar}{\pi}}\ \frac{1}{(p'-p)^2+\delta^2}\ . \qquad (4.19)$$

For this distribution of momenta

$$\langle p\ \rangle = \frac{2\delta^3\hbar}{\pi}\int_{-\infty}^{+\infty} \frac{p'}{((p'-p)^2+\delta^2)^2}\ \frac{dp'}{\hbar} = p$$

$$\langle p^2\rangle = \frac{2\delta^3\hbar}{\pi}\int_{-\infty}^{+\infty}\frac{p'^2}{((p'-p)^2+\delta^2)^2}\frac{dp'}{\hbar} = p^2+\delta^2, \quad \text{i.e.} \quad \Delta p^2 = \delta^2.$$

On the other hand the wavefunction at time t = 0 i.e.

$$\Psi(x, 0) = \frac{1}{\pi}\sqrt{\delta^3\hbar}\int_{-\infty}^{+\infty}\frac{e^{ip'x/\hbar}}{(p'-p)^2+\delta^2}\frac{dp'}{\hbar} =$$

$$\sqrt{\delta^3\hbar}\, e^{ipx/\hbar}\,\frac{1}{\pi}\int_{-\infty}^{+\infty}\frac{e^{i(p'-p)x/\hbar}}{(p'-p)^2+\delta^2}\, d(p'-p)/\hbar =$$

$$= \frac{2\sqrt{\delta^3\hbar}}{\pi\hbar^2}\, e^{\frac{ipx}{\hbar}}\int_0^{\infty}\frac{\cos(p'-p)x/\hbar}{\left(\frac{p'-p}{\hbar}\right)^2+\left(\frac{\delta}{\hbar}\right)^2}\, d(p'-p)/\hbar$$

Hence

$$\Psi(x, 0) = \frac{2}{\pi}\sqrt{\frac{\delta^3}{\hbar^3}}\, e^{ipx/\hbar}\int_0^{\infty}\frac{\cos ux\, du}{u^2+(\delta/\hbar)^2} = \sqrt{\frac{\delta}{\hbar}}\, e^{ipx/\hbar}\, e^{-\frac{\delta|x|}{\hbar}}. \tag{4.20}$$

For this wavefunction

$$\langle x\rangle = 0.$$

$$\langle x^2\rangle = \frac{2\delta}{\hbar}\int_0^{\infty} e^{-2\delta x/\hbar}\, x^2 dx = \frac{\hbar^2}{2\delta^2}$$

i.e.

$$(\Delta x)^2 = \frac{\hbar^2}{2\delta^2}, \quad (\Delta x)^2(\Delta p)^2 = \frac{\hbar^2}{2}, \qquad \Delta x\Delta p = \frac{\hbar}{\sqrt{2}}.$$

The forms of these distributions is given in Figures 4.1 and 4.2 with the appropriate identifications. The symmetry between the A(p') and $\psi(x, 0)$ in this example and the $\psi(x, 0)$ and A(p') in Example 4.1 can be understood in the light of the remarks in Example 4.3.

What is of some interest in this example is that a representation of the delta function is

$$\delta(x) = \frac{1}{\pi}\lim_{\varepsilon\to 0}\frac{1}{\varepsilon}\,\frac{1}{\left(\frac{u}{\varepsilon}\right)^2+1}. \tag{4.21}$$

Thus if one defines

$$A'(p') \equiv \sqrt{\frac{\hbar}{2\pi\delta}}\, A(p') = \frac{\delta\hbar}{\pi} \quad \frac{1}{(p'-p)^2+\delta^2}$$

as

$$\delta \to 0 \qquad A'(p) \to \delta\left(\frac{p'-p}{\hbar}\right).$$

When $\Psi(x, 0)$ is evaluated using $A'(p)$ rather than

$$A(p),\ \Psi(x, 0) = \frac{1}{\sqrt{2\pi}}\, e^{ipx/\hbar}\, e^{-\delta|x|/\hbar}$$

and as $\delta \to 0$ this becomes just a plane wave $\frac{1}{\sqrt{2\pi}} e^{ipx/\hbar}$ as it should since if a delta function is substituted for $A(p')$ in Equation (4.2) it reverts to a plane wave expression.

By a similar analysis one can extract several other delta function representations from the A's used in this Chapter's examples. Some of these are listed in Equation (4.22).

$$\text{(i)} \qquad \delta(u) = \frac{1}{2} \lim_{\varepsilon\to 0} \frac{1}{\varepsilon}\, e^{-|u|/\varepsilon} \qquad \text{(Ex. 4.1)}$$

$$\text{(ii)} \qquad \delta(u) = \lim_{\varepsilon\to 0} \frac{1-|u|/\varepsilon}{\varepsilon} \quad |u| < \varepsilon \qquad \text{(Ex. 4.2)}$$

$$\text{(iii)} \qquad \delta(u) = \frac{1}{\pi} \lim_{\varepsilon\to\infty} \varepsilon\, \frac{\sin^2 \varepsilon u}{(\varepsilon u)^2} \qquad \text{(Ex. 4.3)}$$

$$\text{(iv)} \qquad \delta(u) = \frac{4}{3\pi} \lim_{\varepsilon\to\infty} \frac{\varepsilon \sin^3 \varepsilon u}{(\varepsilon u)^3} \qquad \text{(Ex. 4.5)}$$

$$\text{(v)} \qquad \delta(u) = \frac{3}{2\pi} \lim_{\varepsilon\to\infty} \varepsilon\, \frac{\sin^4 \varepsilon u}{(\varepsilon u)^4} \qquad (4.22)$$

$$\text{(vi)} \qquad \delta(u) = \frac{1}{8} \lim_{\varepsilon\to 0} \frac{1}{\varepsilon}\left(3 - \frac{u^2}{\varepsilon^2}\right) \quad 0 < \frac{|u|}{\varepsilon} < 1$$

$$= \frac{1}{16} \lim_{\varepsilon\to 0} \frac{1}{\varepsilon}\left(3 - \frac{|u|}{\varepsilon}\right)^2 \quad 1 < \frac{|u|}{\varepsilon} < 3\varepsilon \qquad \text{(Ex. 4.5)}$$

$$\text{(vii)} \qquad \delta(u) = \frac{3}{2} \lim_{\varepsilon\to 0} \frac{1}{\varepsilon}\left(1 - \frac{|u|}{\varepsilon}\right)^2 \quad |u| < \varepsilon$$

(viii) $$\delta(u) = \frac{2}{\pi} \lim_{\varepsilon\to 0} \frac{1}{\varepsilon} \frac{1}{(u^2/\varepsilon^2+1)^2} \qquad \text{(Ex. 4.6)}$$

(ix) $$\delta(u) = \frac{1}{4} \lim_{\varepsilon\to 0} \frac{1}{\varepsilon} \left(1+ \frac{|u|}{\varepsilon}\right) e^{-|u|/\varepsilon} \qquad \text{(Ex. 4.6)}$$

Comparing Equation (2.17) with Equation (4.22) (iii) suggests higher powers of $(\varepsilon \sin \varepsilon u)/u$ are also δ functions in appropriate limits.

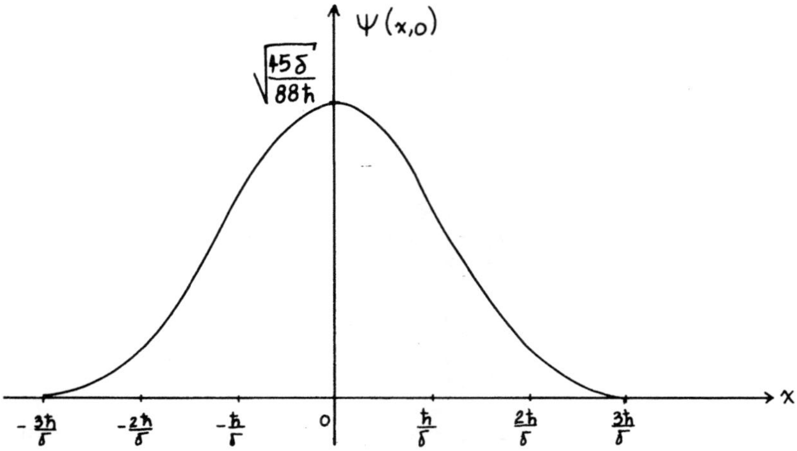

Figure 4.5. Plot of $\Psi(x, 0)$ of Example 4.5.

$$\Psi(x, 0) = \sqrt{\frac{5\delta}{88\hbar}}\, e^{ipx/\hbar} \left[3 - \frac{\delta^2 x^2}{\hbar^2}\right] \qquad 0 < |x| < \frac{\hbar}{\delta}$$

$$= \sqrt{\frac{5\delta}{88\hbar}}\, e^{ipx}\, \frac{1}{2}\left[3 - \frac{\delta|x|}{\hbar}\right]^2 \qquad \frac{\hbar}{\delta} < |x| < \frac{3\hbar}{\delta}$$

$$= 0 \qquad |x| > \frac{3\hbar}{\delta}$$

EXAMPLE 4.5. Consider

$$A(p') = \sqrt{\frac{20\hbar}{11\pi\delta}}\; \frac{\sin^3 (p'-p)/\delta}{((p'-p)/\delta)^3} \tag{4.23}$$

obtain $\Psi(x, 0)$. By straightforward methods one obtains

$$\Psi(x,\ 0) = \frac{1}{\sqrt{2\pi}} \sqrt{\frac{20\hbar}{11\pi\delta}} \int_{-\infty}^{+\infty} \frac{\sin^3 (p'-p)/\delta}{\left(\frac{p'-p}{\delta}\right)^3}\, e^{\frac{ip'x}{\hbar}}\, \frac{dp'x}{\hbar}$$

$$= \sqrt{\frac{10\delta}{11\hbar}}\, e^{ipx/\hbar} \left[\frac{1}{4}\left(3 - \frac{\delta^2 x^2}{\hbar^2}\right)\right] \quad \text{if} \quad 0 < \frac{|x|}{\hbar} < \frac{1}{\delta}$$

$$= \sqrt{\frac{10\delta}{11\hbar}}\, e^{ipx/\hbar} \left[\frac{1}{8}\left(3 - \frac{\delta|x|}{\hbar}\right)^2\right] \quad \text{if} \quad \frac{1}{\delta} < \frac{|x|}{\hbar} < \frac{3}{\delta} \tag{4.24}$$

$$= 0 \qquad |x| > \frac{3}{\delta}$$

Expression (4.24) is plotted in Figure 4.5.

Besides localizing $\Psi(x,\ t)$ the A of expression (4.23) is proportional to another representation of the delta function. In particular

$$\delta(x) = \frac{4}{3\pi} \lim_{u\to\infty} \frac{\varepsilon \sin^3 \varepsilon u}{(\varepsilon u)^3}\ , \tag{4.25}$$

and similarly one may obtain representations of the delta function which involve higher powers of

$$\frac{\sin \varepsilon u}{\varepsilon u}\ , \quad \text{e.g.}\ \frac{3}{2\pi} \lim_{\varepsilon\to\infty} \varepsilon \left(\frac{\sin \varepsilon u}{\varepsilon u}\right)^4 .$$

Also the $\Psi(x,\ 0)$ of Example 4.5 is related to a delta function

$$\delta(x) = \frac{1}{8} \lim_{\delta\to\infty} \delta^3 \left[\left(\frac{3}{\delta^2} - x^2\right)\right] \qquad 0 < |x| < \frac{1}{\delta}$$

$$= \frac{1}{16} \lim_{\delta\to\infty} \delta^3 \left[\frac{1}{2}\left(\frac{3}{\delta} - |x|\right)^2\right] \quad \frac{1}{\delta} < |x| < \frac{3}{\delta} \tag{4.26}$$

$$= 0 \qquad x > \frac{3}{\delta}$$

Higher powers of other delta function expressions for instance of

$$\delta(u) = \frac{1}{\pi} \lim_{\varepsilon\to 0} \frac{1}{\varepsilon} \frac{1}{1+\left(\frac{u}{\varepsilon}\right)^2} \quad \text{and} \quad \delta(u) = \lim_{\varepsilon\to 0} \frac{1-|u|/\varepsilon}{\varepsilon} \quad (|u| < \varepsilon)$$

can also be expressed as delta functions. Thus:

EXAMPLE 4.6. Consider

$$\sqrt{\frac{\hbar\Gamma(2\nu+1)}{\varepsilon\pi^{\frac{1}{2}}\Gamma(2\nu+\frac{1}{2})}}\left[\frac{(p'-p)^2}{\varepsilon^2}+1\right]^{-\nu-\frac{1}{2}} = A(p') \qquad (\nu > -\tfrac{1}{2}) \tag{4.27}$$

a normalized function peaked at $p' = p$.
Find the corresponding $\Psi(x, 0)$.
From standard tables[1]) one obtains

$$\Psi(x, 0) = \sqrt{\frac{2\Gamma(2\nu+1)\varepsilon}{\pi^{\frac{1}{2}}\Gamma(2\nu+\frac{1}{2})\hbar\Gamma(\nu+\frac{1}{2})^2}}\left(\frac{|x|\varepsilon}{2\hbar}\right)^{\nu} K_\nu\left(\frac{\varepsilon|x|}{\hbar}\right) e^{ipx/\hbar}, \tag{4.28}$$

which itself is proportional to a delta function if $\varepsilon \to \infty$.
If $\nu = \frac{1}{2}$ this reduces to Example 4.4. If $\nu = 3/2$,

$$A(p') = \sqrt{\frac{16\hbar}{5\pi\varepsilon}}\left[\frac{(p'-p)^2}{\varepsilon^2}+1\right]^{-2}, \tag{4.29}$$

and

$$\Psi(x, 0) = \sqrt{\frac{2}{5}}\sqrt{\frac{\varepsilon}{\hbar}}\left\{\frac{|x|\varepsilon}{\hbar}+1\right\} e^{-\varepsilon|x|/\hbar} e^{ipx/\hbar}. \tag{4.30}$$

Finally if one replaces $\sqrt{\frac{16\hbar}{5\pi\varepsilon}}$ by $\frac{2\hbar}{\pi\varepsilon}$ in Equation (4.29),

$$\Psi(x, 0) \Rightarrow \sqrt{\frac{1}{2\pi}}\left\{\frac{|x|\varepsilon}{\hbar}+1\right\} e^{-\varepsilon|x|/\hbar} e^{ipx/\hbar},$$

and as $\varepsilon \to 0$

$$\Psi(x, 0) \to \sqrt{\frac{1}{2\pi}}\, e^{ipx/\hbar},$$

implying Equation (4.22) (viii) is a delta function representation. Similar arguments starting from Equation (4.30) lead to the corresponding delta function Equation (4.22) (ix).

EXAMPLE 4.7. Find

$$\Psi(x, t) \quad \text{if} \quad A(p') = \sqrt{\frac{\hbar}{\delta}}\, e^{-|p'-p|/\delta}$$

and assuming the linear approximation:

$$E(p') = E(p) + \left.\frac{dE(p')}{dp'}\right|_{p'=p} (p'-p). \tag{4.31}$$

Using Equation (4.2)

$$\Psi(x, t) = \sqrt{\frac{\hbar}{2\pi\delta}} \int_{p'=-\infty}^{p'=\infty} \frac{dp'}{\hbar} e^{-|p'-p|/\delta + \frac{i}{\hbar}p'x - \frac{iE(p)t}{\hbar} - \frac{i}{\hbar}\frac{dE(p')}{dp'}\Big|_{p'=p}(p'-p)t}.$$

Defining u = p'-p

$$\Psi(x, t) = \sqrt{\frac{1}{2\pi\delta\hbar}}\, e^{\frac{i}{\hbar}\{px-E(p)t\}}\left[\int_0^\infty du\, e^{-u\left\{\frac{1}{\delta}-\frac{ix}{\hbar}+\frac{i}{\hbar}V_g t\right\}} + \int_0^\infty du\, e^{-u\left\{\frac{1}{\delta}+\frac{ix}{\hbar}-\frac{iV_g t}{\hbar}\right\}}\right]$$

$$= \sqrt{\frac{2}{\pi\delta\hbar}}\, e^{\frac{i}{\hbar}\{px-E(p)t\}} \int_0^\infty e^{-\frac{u}{\delta}} \cos\frac{u}{\hbar}(x - V_g t)\, du$$

$$= \sqrt{\frac{2\hbar^3}{\pi\delta^3}} \left(\frac{1}{(x-V_g t)^2 + \frac{\hbar^2}{\delta^2}}\right) e^{\frac{i}{\hbar}(px-E(p)t)}. \tag{4.32}$$

If t = 0 this reduces to the result (4.4) of Example 4.1. Expression (4.32) shows that to the extent the linear approximation (4.31) is valid the wavepacket moves forward at a speed V_g, but its form does not change, i.e. there is no spreading. To get spreading one must keep at least quadratic terms in the expansion for E(p').

EXAMPLE 4.8. Show that

$$\frac{\partial|G|^2}{\partial t} + \frac{\partial}{\partial x}\frac{\hbar}{2mi}\left\{G^*\frac{\partial G}{\partial x} - \frac{\partial G^*}{\partial x}G\right\} = 0, \tag{4.33}$$

where G(x, x', t) is the Green's function for a particular system (i.e. relates Ψ(x, 0) to Ψ(x, t) according to Equation (4.12)).

Since G(x, x', t) satisfies the time-dependent Schrödinger equation (possibly with a potential V, assumed real),

$$\left[-\frac{\hbar^2\partial^2}{2m\partial x^2} + V(x)\right]G = -\frac{\hbar}{i}\frac{\partial}{\partial t}G,$$

and

$$\left[-\frac{\hbar^2\partial^2}{2m\partial x^2} + V(x)\right]G^* = \frac{\hbar}{i}\frac{\partial}{\partial t}G^*.$$

Premultiplying the first of these equations by G* and the second by G one obtains, after subtracting that

$$-\frac{\hbar^2}{2m}\left\{G^*\frac{\partial^2}{\partial x^2}G - G\frac{\partial^2}{\partial x^2}G^*\right\} = -\frac{\hbar}{i}\left\{G^*\frac{\partial}{\partial t}G + G\frac{\partial}{\partial t}G^*\right\} = -\frac{\hbar}{i}\frac{\partial}{\partial t}|G|^2.$$

Hence

$$\frac{\partial}{\partial x}\frac{\hbar}{2mi}\left\{G^*\frac{\partial}{\partial x}G - G\frac{\partial}{\partial x}G^*\right\} + \frac{\partial}{\partial t}|G|^2 = 0.$$

An equation identical to Equation (4.33) is obviously satisfied when G is replaced by Ψ, and G^* by Ψ^*. This is known as the Continuity equation for the current density

$$J = \frac{\hbar}{2mi}\left\{\Psi^*\frac{\partial}{\partial x}\Psi - \Psi\frac{\partial}{\partial x}\Psi^*\right\},$$

and the probability density $\rho = |\Psi|^2$, namely

$$\frac{\partial}{\partial x}J + \frac{\partial}{\partial t}\rho = 0. \tag{4.34}$$

EXAMPLE 4.9. Show that

$$\int_{-\infty}^{+\infty} G^*(x, x', t)\, G(x, x'', t)\, dx = \delta(x' - x''), \tag{4.35}$$

where

$$G(x, x', t) = \sum_{n=0}^{\infty} \phi_n^*(x')\,\phi_n(x)\, e^{-iE_n t/\hbar}, \tag{4.36}$$

(cf. Equation (7.3b)).

Hence show

$$\int_{-\infty}^{+\infty} \delta(x-x')\delta(x-x'')dx = \delta(x'-x''). \tag{4.37}$$

Substituting Equation (4.36) in the integral on the left-hand-side of Equation (4.35) and rearranging the order of summation and integration one obtains:

$$\sum_{n,\, n'} \phi_n(x')\, e^{iE_n t/\hbar}\, \phi_{n'}^*(x'')\, e^{-iE_{n'} t/\hbar} \int_{-\infty}^{+\infty} \phi_n^*(x)\,\phi_{n'}(x)dx =$$

$$\sum_{n,\, n'} \phi_n(x')\, e^{iE_n t/\hbar}\, \phi_{n'}^*(x'')\, e^{-iE_{n'} t/\hbar}\, \delta_{nn'} = \sum_n \phi_n^*(x'')\phi_n(x') =$$

$$= \delta(x'-x''),$$

where one has assumed that the ϕ's satisfy the standard orthonormality and closure conditions, (cf. Equation (2.6)).

Equation (4.37) can be obtained by considering the case t goes to zero in Equation (4.35) since then G(x, x', t) and G*(x, x', t) go to $\delta(x-x')$, while G(x, x", t) goes to $\delta(x-x'')$.
The same result, Equation (4.35) also follows if

$$G(x, x', t) = \int_{-\infty}^{+\infty} \phi_k^*(x')\, \phi_k(x)\, e^{-iE(k)t/\hbar}\, dk, \tag{4.38}$$

(cf. Equation (7.3a), since then Equation (4.35) becomes

$$\int_{-\infty}^{+\infty} dx \int_{-\infty}^{+\infty} dk\, \phi_k^*(x)\, \phi_k(x')\, e^{iE(k)t/\hbar} \int_{-\infty}^{+\infty} dk'\, \phi_{k'}(x)\, \phi_{k'}^*(x'')\, e^{-iE(k')t/\hbar} =$$

$$\int_{-\infty}^{+\infty} dk\, \phi_k(x')\, e^{iE(k)t/\hbar} \int_{-\infty}^{+\infty} dk'\, \phi_{k'}^*(x'')\, e^{-iE(k')t/\hbar} \int_{-\infty}^{+\infty} dx \phi_{k'}(x)\phi_k^*(x) =$$

$$\int_{-\infty}^{+\infty} dk\, \phi_k(x')\, \phi_k^*(x'') = \delta(x'-x''),$$

since

$$\int_{-\infty}^{+\infty} dx\, \phi_{k'}^*(x)\, \phi_k(x) = \delta(k-k'),$$

while

$$\int_{-\infty}^{+\infty} dk\, \phi_k^*(x')\, \phi_k(x) = \delta(x-x'),$$

(cf. Equation (2.7)).

REFERENCE

1. Tables of Integral Transforms, Erdélyi et al., McGraw-Hill (1954), V. 1, p. 11.

TABLE 4.1.
The A(p') and corresponding Ψ(x, 0) discussed in Chapter IV. The A's and ψ'x are normalized according to:

$A(p')$: $\left(\int_{-\infty}^{+\infty} A(p')^2 \frac{dp'}{\hbar} = 1\right)$	$\Psi(x, 0)$: $\left(\int_{-\infty}^{\infty} \lvert\Psi(x, 0)\rvert^2 dx = 1\right)$
$\sqrt{\frac{3\hbar}{2\delta}}\left[1-\frac{\lvert p'-p\rvert}{\delta}\right] \quad \frac{\lvert p-p'\rvert}{\delta} \le 1$ 0 otherwise	$\frac{1}{2}\sqrt{\frac{3\delta}{\pi\hbar}}\, e^{\frac{ipx}{\hbar}} \frac{\sin^2 \frac{\delta x}{2\hbar}}{\left(\frac{\delta x}{2\hbar}\right)^2}$ (Ex. 2)
$\sqrt{\frac{3\hbar}{2\pi\delta}}\, \frac{\sin^2(p'-p)/\delta}{\left(\frac{p'-p)}{\delta}\right)^2}$	$\frac{1}{4}\sqrt{\frac{3\delta}{\hbar}}\, e^{\frac{ipx}{\hbar}} [2-\frac{\lvert x\rvert\delta}{\hbar}] \quad \lvert x\rvert \le \frac{2\hbar}{\delta}$ 0 otherwise (Ex. 3)
$\sqrt{\frac{\hbar}{\delta}}\, e^{-\lvert p'-p\rvert/\delta}$	$\sqrt{\frac{2\delta}{\hbar\pi}}\, e^{\frac{ipx}{\hbar}} \frac{1}{1+\frac{\delta^2 x^2}{\hbar^2}}$ (Ex. 1)
$\sqrt{\frac{2\hbar}{\pi\delta}}\, \frac{1}{1+\left(\frac{p'-p}{\delta}\right)^2}$	$\sqrt{\frac{\delta}{\hbar}}\, e^{\frac{ipx}{\hbar}}\, e^{\frac{-\delta\lvert x\rvert}{\hbar}}$ (Ex. 4)
$\sqrt{\frac{20\hbar}{11\pi\delta}}\, \frac{\sin^3(p'-p)/\delta}{\left(\frac{p'-p}{\delta}\right)^3}$	$\sqrt{\frac{10\delta}{11\hbar}}\, e^{\frac{ipx}{\hbar}} \left(\frac{1}{4}\left(3-\frac{x^2\delta^2}{\hbar^2}\right)\right) \quad 0 < \lvert x\rvert < \frac{\hbar}{\delta}$ $\sqrt{\frac{10\delta}{11\hbar}}\, e^{\frac{ipx}{\hbar}} \left(\frac{1}{8}\left(3-\frac{\lvert x\rvert\delta}{\hbar}\right)^2\right) \quad \frac{\hbar}{\delta} < \lvert x\rvert < \frac{3\hbar}{\delta}$ 0 otherwise (Ex. 5)
$\sqrt{\frac{\hbar\Gamma(2\nu+1)}{\delta\pi^{\frac{1}{2}}\Gamma(2\nu+\frac{1}{2})}}\left[\left(\frac{p'-p}{\delta}\right)^2+1\right]^{-\nu-\frac{1}{2}}$ $\nu>-\frac{1}{2}$	$\sqrt{\frac{2\Gamma(2\nu+1)\delta}{\pi^{\frac{1}{2}}\Gamma(2\nu+\frac{1}{2})\hbar\Gamma^2(\nu+\frac{1}{2})}}\, e^{\frac{ipx}{\hbar}} \left(\frac{\lvert x\rvert\delta}{2\hbar}\right)^{\nu} K_{\nu}\left(\frac{\delta\lvert x\rvert}{\hbar}\right)$ (Ex. 6)
$\sqrt{\frac{16\hbar}{5\pi\delta}}\left[\left(\frac{p'-p}{\delta}\right)^2 + 1\right]^{-2}$	$\sqrt{\frac{2\delta}{5\hbar}}\, e^{\frac{ipx}{\hbar}} \left[\frac{\delta\lvert x\rvert}{\hbar} + 1\right] e^{-\delta\lvert x\rvert/\hbar}$ (Ex. 6. $\nu = \frac{3}{2}$)

CHAPTER 5

Uncertainty Principle and Ground State Energies of Quantum Mechanical Systems

Consider a particle moving subject to a potential $V(x) = A|x|^n$, $-\infty<x<\infty$. Classically the particle's energy is:

$$E = p^2/2m + A|x|^n. \tag{5.1}$$

But $\Delta x\, \Delta p \sim \hbar$ (cf. Chapter 4) i.e. $\Delta p \sim \frac{\hbar}{\Delta x}$ and one expects $p^2 > (\Delta p)^2 = \hbar^2/(\Delta x)^2$ since generally p is expected to be at least of the order of Δp.

Hence

$$E \geq \frac{(\Delta p)^2}{2m} + A|x|^n = \frac{\hbar^2}{2m(\Delta x)^2} + A|x|^n.$$

Requiring that Δx is of the order of 2x one obtains

$$E = \frac{\hbar^2}{2m(\Delta x)^2} + A\left(\frac{\Delta x}{2}\right)^n,$$

where Δx is assumed greater than zero.
The choice of Δx which minimizes E is such that

$$\frac{\partial E}{\partial \Delta x} = 0 \qquad \text{i.e. } 0 = -\frac{\hbar^2}{m(\Delta x)^3} + n\,\frac{A}{2^n}(\Delta x)^{n-1}$$

$$\text{implying } (\Delta x)^{n+2} = \frac{\hbar^2 2^n}{nmA}.$$

For this choice of Δx

$$E \approx \frac{1}{(\Delta x)^2}\left\{\frac{\hbar^2}{2m} + \frac{A}{2^n}(\Delta x)^{n+2}\right\} = \frac{\hbar^2}{2mn}\,\frac{(n+2)}{(\Delta x)^2}$$

$$E \geq \frac{n+2}{2^{\frac{3n+2}{n+2}}} \left\{ \frac{\hbar^2 A^{2/n}}{nm} \right\}^{\frac{n}{n+2}} \qquad (5.2)$$

EXAMPLE 5.1. Consider the case n = 2 $-\infty < x < \infty$.

$$E = \frac{4}{2^2}\left(\frac{\hbar^2 A}{2m}\right)^{\frac{1}{2}} = \frac{1}{\sqrt{2}}\left(\frac{\hbar^2 A}{m}\right)^{\frac{1}{2}} .$$

The exact ground state energy of this system is also

$$\frac{1}{\sqrt{2}}\left(\frac{\hbar^2 A}{m}\right)^{\frac{1}{2}} .$$

EXAMPLE 5.2. Consider the case n = 1 $-\infty < x < \infty$.

$$E = \frac{3}{2^{5/3}}\left\{\frac{\hbar^2 A^2}{m}\right\}^{1/3} \doteq 0.94 \left\{\frac{\hbar^2 A^2}{m}\right\}^{1/3} .$$

This can be compared to the ground state energy of this system (for large arguments of the relevant Airy function (cf. Equation (3.31)) namely 0.89 $\{\hbar^2 A^2/m\}^{1/3}$. Meanwhile a variational calculation (Example 10.3) yields

$$E_g \leq 0.81 \left\{\frac{\hbar^2 A^2}{m}\right\}^{1/3} .$$

Example 5.3. Consider the case n = -1

$$E = 2\left\{\frac{\hbar^2 A^{-2}}{-m}\right\}^{-1} = -\frac{2mA^2}{\hbar^2} .$$

This agrees exactly with the ground state energy of the svstem (cf. Equation (3.32)).

If $E = p^2/2m + Ax^n$ only for $x > 0$ the above derivation must be modified in that $\Delta x \sim x$ and

$$E = \frac{n+2}{2}\left\{\frac{\hbar^2 A^{2/n}}{nm}\right\}^{\frac{n}{n+2}} \qquad (5.3)$$

EXAMPLE 5.4. If n = 2 $x \geq 0$

$$E = \frac{4}{2}\left\{\frac{\hbar^2 A}{2m}\right\}^{\frac{1}{2}} = \sqrt{2}\left\{\frac{\hbar^2 A}{m}\right\}^{\frac{1}{2}} ,$$

compared to the exact ground state energy in this case (cf. Example 8.1)

$$= \sqrt{4.5}\left\{\frac{\hbar^2 A}{m}\right\}^{\frac{1}{2}} .$$

EXAMPLE 5.5. If n = -1 $x \geq 0$,

$$E = \frac{3}{2}\left\{\frac{\hbar^2 A^2}{m}\right\}^{1/3} ,$$

which can be compared to the variational calculation for this problem (cf. Example 10.2) $E_{ground} < 1.86\ [A^2\hbar^2/m]^{1/3}$ and the lowest energy of this system for large arguments of the relevant Airy function, namely $1.84\ (A^2\hbar^2/m)^{1/3}$ (cf. Equation (3.18)).

EXAMPLE 5.6. If n =-1, $x \geq 0$,

$$E = \frac{1}{2}\left\{\frac{\hbar^2 A^{-2}}{-m}\right\}^{-1} = -\frac{mA^2}{2\hbar^2} ,$$

which agrees exactly with the ground state energy of the system (cf. Equation (3.23)).

One should emphasize that expressions (5.2) and (5.3) are very rough estimates of the ground state energy of quantum mechanical systems, but are nonetheless convenient if one is interested in a result which is of the right order of magnitude.

EXAMPLE 5.7. Consider a particle moving in the attractive potential

$$V(x) = -\frac{|V_0|\varepsilon}{\pi(x^2+\varepsilon^2)} ,$$

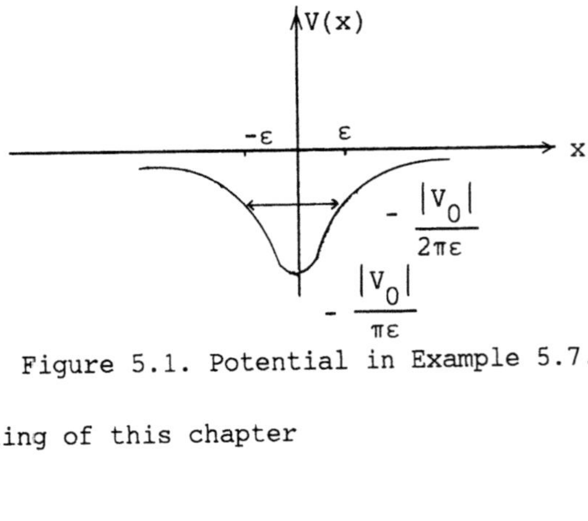

Figure 5.1. Potential in Example 5.7.

Using the arguments at the beginning of this chapter

$$E = \frac{\hbar^2}{2m(\Delta x)^2} - \frac{|V_0|\varepsilon}{\pi\left(\left(\frac{\Delta x}{2}\right)^2 + \varepsilon^2\right)} ,$$

and

$$\frac{\partial E}{\partial x} = - \frac{\hbar^2}{m(\Delta x)^3} + \frac{|V_0|\varepsilon\Delta x}{2\pi\left(\left(\frac{\Delta x}{2}\right)^2 + \varepsilon^2\right)^2} .$$

The choice of Δx which minimizes E is thus

$$(\Delta x)^2 = \frac{4\alpha\varepsilon^2}{4-\alpha} \quad \text{where} \quad \alpha = \sqrt{\frac{2\pi\hbar^2}{m|V_0|\varepsilon}} .$$

Therefore

$$E_{min} = \left[- \frac{|V_0|}{\pi\varepsilon} + \hbar\sqrt{\frac{|V_0|}{2\pi m\varepsilon^3}} - \frac{\hbar^2}{8m\varepsilon^2}\right] .$$

As ε tends to ∞,

$$E_{min} \to \left[- \frac{|V_0|}{\pi\varepsilon} + \hbar\sqrt{\frac{|V_0|}{2\pi m\varepsilon^3}}\right] .$$

This is consistent with Example 5.1 since then

$$V(x) \sim C + Ax^2 \quad \text{where} \quad C = - \frac{|V_0|}{\pi\varepsilon} \quad \text{and} \quad A = \frac{|V_0|}{\pi\varepsilon^3} .$$

Free Particles Incident on Potentials, Time Delay, Phase Shifts and the Born Approximation

When quantum mechanical particles are incident on a potential one is in the first instance interested in the fraction transmitted through the potential and the fraction reflected by it. One therefore calculates the probability of reflection and the probability of transmission.

In detail if one writes for a particle of energy

$$E = \frac{p^2}{2m} = \frac{\hbar^2 k^2}{2m}$$

that the wavefunction on the left side of a one dimensional potential is:

$$\psi_L(x) = Ae^{ikx} + Be^{-ikx}, \tag{6.1}$$

and that the wavefunction on the right side is

$$\psi_R(x) = Fe^{ikx}, \tag{6.2}$$

this choice implies the particle is incident on the potential from the left. Additionally

$R = B/A$ is the reflection amplitude, with $|R|^2$ the reflection probability and $T = F/A$ is the transmission amplitude, with $|T|^2$ the transmission probability. (6.3)

The poles of the transmission amplitude correspond to the allowed bound states for that particular potential, and continuity considerations require

$$|B|^2 + |F|^2 = |A|^2 \quad \text{i.e.} \quad |R|^2 + |T|^2 = 1. \tag{6.4}$$

In addition an incident wavepacket can be written (cf. Chapter 4) as

$$\Psi_{in}(x, t) = \frac{1}{\sqrt{2\pi}} \int A(k') \, e^{i(k'x-E(k')t/\hbar)} \, dk', \tag{6.5}$$

and based on this the reflected and transmitted wavepackets are

$$\Psi_{ref}(x, t) = \frac{1}{\sqrt{2\pi}} \int R(k')A(k') \, e^{-i(k'x+E(k')t/\hbar)} \, dk' \tag{6.6}$$

and

$$\psi_{trans}(x, t) = \frac{1}{\sqrt{2\pi}} \int T(k')A(k') \, e^{i(k'x-E(k')t/\hbar} \, dk' \tag{6.7}$$

respectively.

Using standard procedures (cf. Equation (4.14)) one can then obtain from expressions (6.5), (6.6), and (6.7) the time required for reflection off (the so called time delay) and transmission through a particular potential.

EXAMPLE 6.1. Discuss the problem of particles incident on a potential $V(x) = V_0\delta(x)$. If

$$\psi_L = Ae^{ikx} + Be^{-ikx}$$
$$\psi_R = Fe^{ikx} \tag{6.8}$$

continuity of the wavefunction at $x = 0$ implies:

$$A + B = F. \tag{6.9}$$

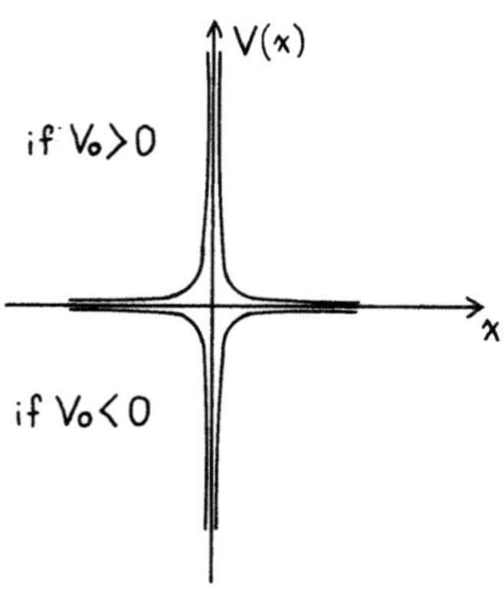

Figure 6.1. Potentials in Example 6.1 if $V_0 < 0$ or $V_0 > 0$.

This potential is discontinuous at x = 0. Thus, using the Schrödinger equation

$$-\frac{\hbar^2}{2m}\frac{d^2}{dx^2}\psi(x) + V_0\delta(x)\,\psi(x) = E\psi(x),$$

and integrating across x = 0

$$-\frac{\hbar^2}{2m}\int_{-\varepsilon}^{\varepsilon} d\left(\frac{d\psi}{dx}\right) + \int_{-\varepsilon}^{\varepsilon} V_0\delta(x)\psi(x)\,dx = E\int_{-\varepsilon}^{\varepsilon}\psi(x)\,dx.$$

In the limit as $\varepsilon \to 0$ this becomes

$$\lim_{\varepsilon\to 0}\left\{\frac{d\psi}{dx}\Bigg|_{x=\varepsilon} - \frac{d\psi}{dx}\Bigg|_{x=-\varepsilon}\right\} = \frac{2mV_0}{\hbar^2}\psi(0) , \tag{6.10}$$

i.e.

$$ik(F-A+B) = \frac{2mV_0}{\hbar^2} F. \tag{6.11}$$

Solving Equations (6.9) and (6.11) simultaneously yields:

$$R = \frac{1}{i\frac{k\hbar^2}{mV_0}-1} = -\frac{e^{i\delta_r}}{\sqrt{1+\frac{k^2\hbar^4}{m^2V_0^2}}} , \quad \delta_r = \tan^{-1}\frac{k\hbar^2}{mV_0} \tag{6.12}$$

$$T = \frac{\frac{ik\hbar^2}{mV_0}}{\frac{ik\hbar^2}{mV_0}-1} = -\frac{\frac{k\hbar^2}{mV_0}e^{i(\delta_r+\pi/2)}}{\sqrt{1+\frac{k^2\hbar^4}{m^2V_0^2}}} , \quad \delta_t = -\cot^{-1}\frac{k\hbar^2}{mV_0} = \delta_r+\frac{\pi}{2}. \tag{6.13}$$

In this case

$$|R|^2 = \frac{1}{\frac{k^2\hbar^4}{m^2V_0^2}+1}$$

$$|T|^2 = \frac{\dfrac{k^2\hbar^4}{m^2V_0^2}}{\dfrac{k^2\hbar^4}{m^2V_0^2}+1} \tag{6.14}$$

and as expected $R^2 + T^2 = 1$, consistent with Equation (6.5).

The poles of the transmission amplitude satisfy $(ik\hbar^2)/(mV_0) = 1$, i.e.

$$\frac{k^2\hbar^2}{2m} = -\frac{mV_0^2}{2\hbar^2} = E \ldots . \tag{6.15}$$

There is only one pole here and this is physically meaningful when $V_0 < 0$. To get reflection and transmission times one must construct wavepackets

$$\Psi_{inc}(x, t) = \frac{1}{\sqrt{2\pi}} \int dk' A(k')\, e^{i(k'x - \frac{\hbar k'^2 t}{2m})} \approx e^{i\phi}\, \Psi_{in}(x - v_g t, 0)$$

$$\Psi_{ref}(x, t) = -\frac{1}{\sqrt{2\pi}} \int dk' \frac{A(k')}{\sqrt{1 + \dfrac{k'^2\hbar^4}{m^2V_0^2}}}\, e^{-i(k'x + \frac{\hbar k^2}{2m}t - \delta_r)} \approx e^{i\eta}\Psi_{ref}\left(x + v_g t - \left.\frac{d\delta_t}{dk'}\right|_{k'=k}, 0\right)$$

$$\Psi_{trans}(x, t) = \frac{-1}{\sqrt{2\pi}} \int dk' \frac{A(k')\dfrac{k'\hbar^2}{mV_0}}{\sqrt{1 + \dfrac{k'^2\hbar^4}{m^2V_0^2}}}\, e^{i(k'x - \frac{\hbar k'^2 t}{2m}\ \delta_t)} \approx e^{i\theta}\Psi_{trans}\left(x - v_g t + \left.\frac{d\delta_t}{dk'}\right|_{k'=k}, 0\right) \tag{6.16}$$

where

$$v_g = \left.\frac{\hbar k'}{m}\right|_{k'=k}$$

and ϕ, η and θ are phases. If at $t = 0$, $x = 0$ for the incident wavepacket. in the reflected wavepacket

$$x = 0 \text{ at } t = \left.\frac{1}{v_g}\frac{d\delta_r}{dk'}\right|_{k'=k}$$

while in the transmitted wavepacket

$$x = 0 \text{ at } t = \frac{1}{v_g} \left. \frac{d\delta_t}{dk'} \right|_{k'=k} .$$

Thus

$$t_{ref} = \frac{1}{v_g} \frac{\frac{\hbar^2}{mV_0}}{1 + \frac{k^2\hbar^4}{m^2V_0^2}} = t_{trans} \ldots \qquad (6.17)$$

$$= \frac{\hbar^2}{mv_gV_0} \left\{ \frac{1}{1 + \frac{k^2\hbar^4}{m^2V_0^2}} \right\}$$

as $k \to \infty$ t_{ref}, $t_{trans} \to 0$, the classical result.

EXAMPLE 6.2. Discuss the problem of particles incident on the potential:

$$V(x) = V_0\delta(x+a) \qquad x < 0$$

$$= \infty \qquad x > 0.$$

If

$$\psi_L = Ae^{ik(x+a)} + Be^{-ik(x+a)} \qquad x < -a,$$

$$\psi_R = C\sin kx \quad -a < x < 0 \qquad (\text{consistent with } \psi(0) = 0),$$

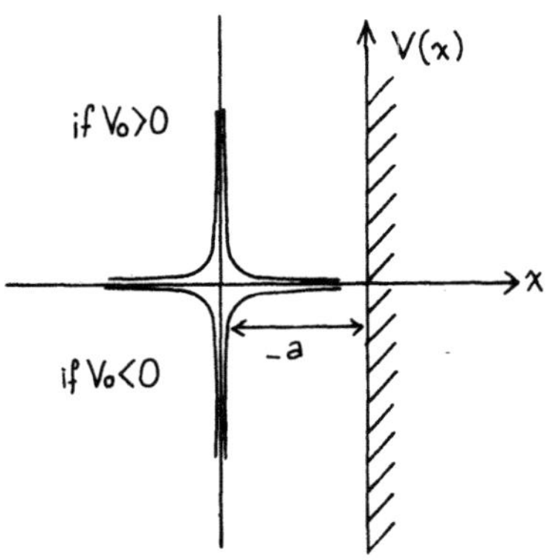

Figure 6.2. Potential in Example 6.2.

continuity of the wavefunction implies

$$A + B = - C \sin ka. \tag{6.18}$$

The discontinuity of the derivative of ψ at $x = -a$ (see Equation (6.10)) implies

$$kC \cos ka - ik(A-B) = \frac{2mV_0}{\hbar^2} (A+B) . \tag{6.19}$$

$$R = \frac{B}{A} = \frac{\dfrac{-2mV_0}{\hbar^2} - k \cot ka - ik}{\dfrac{2mV_0}{\hbar^2} + k \cot ka - ik} = -e^{i\delta_r},$$

$$\delta_r = 2 \cot^{-1}\left(\cot ka + \frac{2mV_0}{\hbar^2 k}\right) . \tag{6.20}$$

One notes $|R|^2 = 1$. In fact B/A in this case is both the reflection amplitude and the transmission amplitude since there is no transmission beyond $x = 0$.

The allowed negative energies satisfy the equation:

$$\frac{2mV_0}{\hbar^2} + k\cot ka = ik. \tag{6.21}$$

Defining

$i\kappa = k = \sqrt{\dfrac{2mE}{\hbar^2}} = i \sqrt{\dfrac{2m|E|}{\hbar^2}}$, Equation (6.21) becomes

$$\frac{2mV_0}{\hbar^2} + \kappa \coth \kappa a + \kappa = 0,$$

for the allowed bound states. Also the particle is reflected from $x = -a$ at

$$\tau_{ref} = \frac{1}{v_g} \left.\frac{d\delta_r(k')}{dk'}\right|_{k'=k} = \frac{2a}{v_g} \left(\frac{\csc^2 ka + \dfrac{2mV_0}{\hbar^2 k^2 a}}{1 + (\cot ka + \dfrac{2mV_0}{\hbar^2 k})^2}\right)_{k'=k} .$$

as k becomes very large

$$\tau_{ref} \to \frac{2a}{v_g} \left\{\frac{\csc^2 ka}{1+\cot^2 ka}\right\} = \frac{2a}{v_g} ,$$ the classical value.

In three dimensional systems where particles of mass m are incident on potentials V(r) of finite range a, the radial wavefunction for r > a can be written:

$$\frac{u_\ell(r)}{r} = R_\ell(r) = A_\ell(j_\ell(kr) - \tan \delta_\ell(k)\ \eta_\ell(kr)) \tag{6.22}$$

where j_ℓ, n_ℓ are the usual spherical Bessel and Neumann functions and $\delta_\ell(k)$ is the "phase shift" for a particular "wave" or angular momentum ℓ at incident energy $E = \hbar^2k^2/2m$.
In terms of these phase shifts a standard derivation shows the scattering amplitude, the three dimensional analogue of the transmission amplitude of Equation (6.3) is:

$$f_k(\theta) = \frac{1}{k} \sum_{\ell=0}^{\infty} e^{i\delta_\ell(k)} \sin \delta_\ell(k) P_\ell(\cos\theta), \tag{6.23}$$

while the (scattering) cross section $\sigma(k) = \int |f_k(\theta)|^2 d\Omega$

$$= \frac{4\pi}{k^2} \sum_{\ell=0}^{\infty} (2\ell+1) \sin^2 \delta_\ell(k). \tag{6.24}$$

As in the one-dimensional case the poles in the scattering amplitude for a particular potential give its allowed negative energies.

The phase shifts $\delta_\ell(k)$ may be obtained exactly, by requiring that the wavefunction (and its derivative if V has no infinite discontinuities) in the region where there is a potential, match smoothly onto the external wavefunction Equation (6.22) (and its derivative) at r = a.

Alternatively at high incident particle energies or for weak potentials one may have recourse to the partial wave Born approximation:

$$\tan \delta_\ell(k) = - \frac{2mk}{\hbar^2} \int_0^\infty j_\ell^2(kr) V(r) r^2\, dr, \tag{6.25}$$

or the (first) Born approximation which for central (ℓ independent) potentials reduces to

$$f_B(k') = - \frac{2m}{\hbar^2} \int_0^\infty \frac{\sin k'r'}{k'r'} V(r') r'^2\, dr', \tag{6.26}$$

(with $k' = 2k \sin\theta/2$), where the differential cross section is:

$$\frac{d\sigma}{d\Omega} = |f(k')|^2 , \tag{6.27}$$

while the cross section σ is

$$\sigma = \int |f(k)|^2 d\Omega \simeq \int |f_B(k')|^2 d\Omega. \tag{6.28}$$

EXAMPLE 6.3. Consider a particle scattered off the three-dimensional potential $V(r) = V_0\delta(r-a)$

$$u_\ell(r) = B_\ell \, rkj_\ell \, (kr) \qquad 0 < r < a$$

$$= A_\ell \, rk \, (j_\ell \, (kr) - \tan \delta_\ell \, (k) \, n_\ell \, (kr)) \quad a < r < \infty.$$

For angular momentum $\ell = 0$ if one matches the two pieces of the wavefunction at $r = a$

$$B_0 \sin ka = \frac{A_0}{\cos \delta_0} \sin(ka+\delta_0), \tag{6.29}$$

while the discontinuity of the derivative of $u_0(r)$ at $r = a$ (see Equation (6.10)) requires:

$$\frac{kA_0}{\cos \delta_0} \cos(ka+\delta_0) - B_0 k \cos ka = \frac{2mV_0}{\hbar^2} B_0 \sin ka. \tag{6.30}$$

The ratio of expression (6.30) to expression (6.29) is:

$$\cot(ka+\delta_0) - \cot ka = \frac{2mV_0}{k\hbar^2},$$

i.e.

$$\cot \delta_0 = -(\cot ka + \frac{\hbar^2 k}{2mV_0} \csc^2 ka)$$

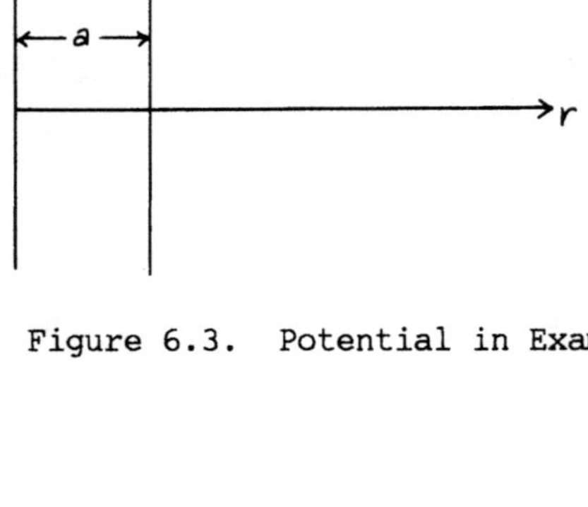

Figure 6.3. Potential in Example 6.3.

or

$$\sin\delta_0 = \frac{-1}{\csc ka\sqrt{1+2A\cot ka + A^2\csc^2 ka}} \;,$$

where

$$A = \frac{\hbar^2 k}{2mV_0} \;.$$

The $\ell = 0$ contribution to the scattering amplitude is

$$\frac{1}{k}\, e^{i\delta_0(k)} \quad \sin\delta_0(k),$$

which has poles when

$$1 + 2\left(\frac{k\hbar^2}{2mV_0}\right)\cot ka + \left(\frac{k\hbar^2}{2mV_0}\right)^2 \csc^2 ka = 0$$

i.e. when

$$\left(1 + \frac{k\hbar^2}{2mV_0}\cot ka\right)^2 = -\left(\frac{k\hbar^2}{2mV_0}\right)^2$$

which is in agreement with the result of Equation (6.21) as it must be.

For higher ℓ's different (additional) bound states result. The $\ell = 0$ contribution to the scattering cross section is:

$$\frac{4\pi}{k^2\csc^2 ka\left(1+\frac{k\hbar^2}{mV_0}\cot ka + \left(\frac{k\hbar^2}{2mV_0}\right)^2\csc^2 ka\right)} \;.$$

If $a \to 0$ the $\delta_0(k)$ goes to zero and with it f and σ.

One notes that $\delta_{\ell=0}$ depends on the sign of V_0.

As concerns the phase shift for $\ell = 0$ if k is large, i.e. A is large (where $A = (\hbar^2 k)/(2mV_0)$),

$$\sin\delta_0 \simeq \frac{-1}{\csc ka\; A\; \csc ka} \simeq -\frac{2mV_0\sin^2 ka}{\hbar^2 k} \;, \tag{6.31}$$

while the partial wave Born approximation $\ell=0$ phase shift may be easily evaluated:

$$\tan\delta_0(k) \simeq -\frac{2mk}{\hbar^2}\, j_0^2(ka)V_0a^2 = -\frac{2mV_0\sin^2 ka}{\hbar^2 k} \;,$$

in agreement with Equation (6.31) as it should be.

The first Born approximation for this potential (Equation (6.26)), can also be evaluated:

$$f_B = -\frac{2m}{\hbar^2}\int \frac{\sin k'r'}{k'r'} r'^2 V_0 \delta(r'-a)\, dr' = -\frac{2mV_0 a}{\hbar^2 k'} \sin k'a,$$

yielding a differential cross section (Equation (6.27)),

$$\frac{d\sigma}{d\Omega} = \frac{m^2 V_0^2 a^2}{\hbar^4 k^2 \sin^2 \theta/2} \sin^2 (2a\, k \sin \theta/2).$$

EXAMPLE 6.4. Consider a particle scattered off the three dimensional potential

$$V(r) = \hbar^2/mr^2 \qquad 0 < r < a$$

$$V(r) = 0 \qquad r > a,$$

for $\ell = 0$ (S wave) scattering.

For

$$r > a \qquad u_0(r) = \frac{A_0}{\cos \delta_0} \sin(kr + \delta_0).$$

For $0 < r < a$ $u_0(r)$ is the solution of the equation

$$\left\{-\frac{\hbar^2}{2m}\frac{d^2}{dr^2} + \frac{2\hbar^2}{2mr^2} - E\right\} u_0(r) = 0. \qquad (6.32)$$

The solution of Equation (6.32) which does not diverge at the origin is $B_0\, r\, j_1\, (kr)$. Hence one can easily find the $\ell = 0$ phase shift for this system by requiring continuity of the wavefunction and its derivative at $r = a$:

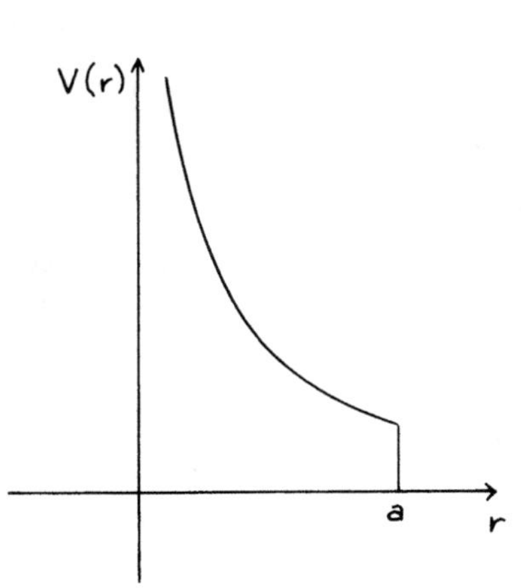

Figure 6.4. Potential in Example 6.4.

$$B_0 \, aj_1(ka) = \frac{A_0}{\cos \delta_0} \sin(ka + \delta_0)$$

$$B_0 \frac{d}{dr} (rj_1(kr))\Big|_{r=a} = \frac{A_0 k}{\cos \delta_0} \cos (ka+\delta_0).$$

Thus

$$\frac{1}{k} \tan(ka+\delta_0) = \frac{aj_1(ka)}{\frac{d}{dr}(rj_1(kr))\Big|_{r=a}}$$

where

$$j_1(\rho) = \frac{\sin \rho}{\rho^2} - \frac{\cos \rho}{\rho}$$

i.e.

$$\delta_0(k) = \cot^{-1} \frac{d}{d\rho} \ln(\rho j_1 (\rho))\Big|_{\rho=ka} -ka.$$

Evaluating this one obtains an explicit expression for $\delta_0(k, a)$

$$\tan \delta_0(k, a) = - ka \left\{\frac{\tan ka + \cot ka - (\frac{1}{ka})^2 \tan ka}{\tan ka + \cot ka - 1/ka}\right\} .$$

As $k \to \infty$ $\tan \delta_0 \to - ka$, the infinite barrier result since for an infinitely repulsive barrier of radius a,

$$0 = \sin(ka + \delta_0), \quad \text{i.e.} \quad \delta_0 = - ka. \tag{6.33}$$

If a is very short range (or k is small) such that ka is small,

$$\tan \delta_0 \simeq - ka\left\{\frac{ka+\left(\frac{1}{ka} - \frac{ka}{3}\right) - \left(\frac{1}{ka}\right)^2\left(ka + \frac{(ka)^3}{3}\right)}{ka + \frac{1}{ka} - \frac{ka}{3} - \frac{1}{ka}}\right\} = - \frac{ka}{2} .$$

In the Born approximation the total differential cross section for this potential may be obtained by first evaluating the scattering amplitude:

$$f_B(k') = - \frac{2m}{\hbar^2} \frac{\hbar^2}{m} \int_0^a \frac{\sin k'r'}{k'r'} dr'.$$

As $a \to \infty$ this becomes

$$- \frac{\pi}{k'} = - \frac{\pi}{2k \sin \theta/2} .$$

The cross section in this case ($a \to \infty$) for scattering between $\theta = \theta_1$ and $\theta = \theta_2$ is

$$\frac{\pi^2}{4k^2} 2\pi \int_{\theta_1}^{\theta_2} \frac{\sin\theta\, d\theta}{\sin^2\theta/2} = \frac{\pi^2}{4k^2} 2\pi \int_{\theta_1}^{\theta_2} \frac{2 \sin\theta/2 \cos\theta/2\ 2\, d\,\theta/2}{\sin^2\theta/2}$$

$$= \frac{2\pi^3}{k^2}\left(\ln \sin\theta_2/2 - \ln \sin\theta_1/2\right) = \frac{2\pi^3}{k^2} \ln\left(\frac{\sin\theta_2/2}{\sin\theta_1/2}\right) .$$

EXAMPLE 6.5. Consider a particle scattered off the potential

$$V(r) = -|V_0| + \frac{1}{2} m\omega^2 r^2 \qquad 0 < r < a$$

$$= 0 \qquad r > a$$

which is purely attractive or partly attractive and partly repulsive depending on the values of V_0, ω and a. (See Figure 6.5.)

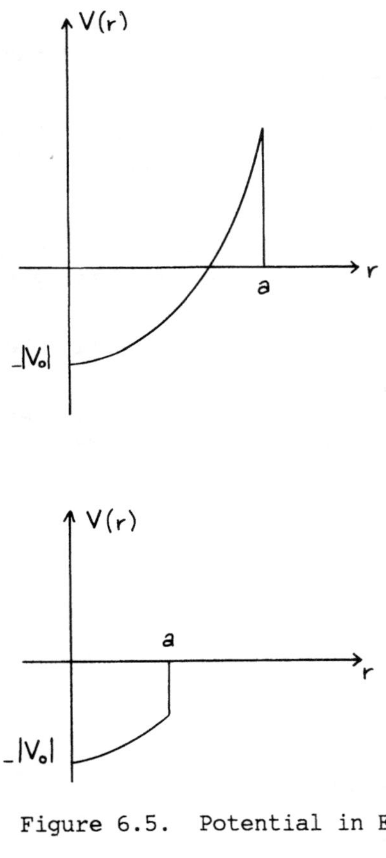

Figure 6.5. Potential in Example 6.5.

Consider in particular the special case

$$E = -|V_0| + \frac{3}{2}\hbar\omega = \frac{\hbar^2 k^2}{2m} .$$

At this energy

$$u_0(r) = Nre^{-m\omega r^2/2\hbar} \qquad 0 < r < a.$$

Matching the internal and external wavefunctions at r = a yields

$$\frac{Na\, e^{-m\omega a^2/2\hbar}}{N\, e^{-m\omega a^2/2\hbar} - N\,\frac{a^2 m\omega}{\hbar}\, e^{-m\omega a^2/2\hbar}} = \frac{1}{k}\tan(ka+\delta_0) = \frac{a}{1-\frac{a^2 m\omega}{\hbar}} .$$

Hence

$$\tan(ka+\delta_0) = \frac{ka}{1-\frac{a^2 m\omega}{\hbar}} ,$$

i.e.

$$\tan\delta_0(k, a) = \frac{ka + \frac{m\omega a^2}{\hbar}\tan ka - \tan ka}{1 - \frac{m\omega}{\hbar}a^2 + ka\tan ka} = \frac{ka+\frac{2ma^2}{3\hbar^2}(E+|V_0|)\tan ka-\tan ka}{1 - \frac{2ma^2}{3\hbar^2}(E+|V_0|)+ka\tan ka} . \tag{6.34}$$

If ω is large the potential looks like an infinitely repulsive barrier of radius r = a and the scattering takes place at high energy since $E = -|V_0| + 3/2\ \hbar\omega$.

In this case $\tan\delta_0 \approx -\tan ka$ i.e. $\delta_0 = -ka$, exactly the infinite repulsive barrier result Equation (6.33).

If ω is small this problem reduces to scattering at low energy off a square well of depth $|V_0|$ (since small ω implies small E). For square well scattering

$$\tan\delta_0 = \frac{\frac{k}{K}\tan Ka - \tan ka}{1+\frac{k}{K}\tan Ka\tan ka}$$

where

$$K = \sqrt{\frac{2m}{\hbar^2}\left\{E + |V_0|\right\}}$$

and the potential $V = -|V_0|$ has a range $0 \le r \le a$.

At low energy

$$\frac{k}{K}\tan ka \approx \frac{k}{K}\left\{Ka + \frac{(Ka)^3}{3}\right\} = ka + \frac{kK^2a^3}{3} = ka + \frac{ka^3 2m}{3\hbar^2}\left\{E + |V_0|\right\},$$

hence for a square well scattering at low energy

$$\tan\delta_0 = \frac{ka + \frac{2ka^3m}{3\hbar^2}\left\{E_0 + |V_0|\right\} - ka}{1+k^2a^2 + \frac{2k^2a^4m}{3\hbar^2}\left\{E_0 + |V_0|\right\}} \sim \frac{2ka^3m}{3\hbar^2}\left\{E + |V_0|\right\}. \quad (6.35)$$

Equation (6.34) agrees with Equation (6.35) in this limit.

The total scattering amplitude for this potential in the Born approximation is easily obtained:

$$f(k') = -\frac{2m}{\hbar^2}\int_0^a \frac{\sin k'r'}{k'r'}\left(-|V_0| + \tfrac{1}{2}m\omega^2r'^2\right)r'^2dr'$$

$$= \frac{2m|V_0|}{\hbar^2k'^3}\int_0^{k'a} k'r' \sin k'r' \, d(k'r') - \frac{m^2\omega^2}{k'^5\hbar^2}\int_0^{k'a} (k'r')^3 \sin k'r' \, d(k'r')$$

$$= \frac{2m|V_0|}{\hbar^2k'^3}\left\{\sin k'a - k'a\cos k'a\right\} - \frac{m^2\omega^2}{k'^5\hbar^2}\left\{(3(k'a)^2-6)\sin k'a + k'a(6-(k'a)^2)\cos k'a\right\}.$$

EXAMPLE 6.6. Evaluate the differential and total scattering cross section for the potential $V(r) = V_0\, e^{-r^2/2a^2}$ in the Born approximation.

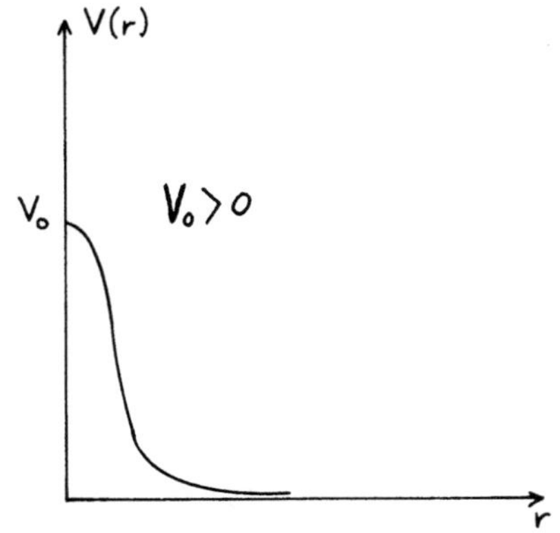

Figure 6.6. Potential in Example 6.6.

$$f_B(k') = -\frac{2mV_0}{\hbar^2}\int_0^\infty \frac{\sin k'r'}{k'r'}\, e^{-r'^2/2a^2}\, r'^2 dr'$$

$$= -\frac{mV_0}{2\hbar^2 k'}\int_{r'=-\infty}^{r'=\infty} \sin k'r'\, e^{-r'^2/2a^2}\, dr'^2$$

$$= -\frac{mV_0}{2\hbar^2 k'}\,\mathrm{Im}\int_{-\infty}^{+\infty} e^{ik'r'-r'^2/2a^2}\, dr'^2$$

$$= -\frac{mV_0}{2\hbar^2 k'}\, e^{-\frac{k'^2a^2}{2}}\,\mathrm{Im}\int_{-\infty}^{+\infty} e^{-\frac{1}{2a^2}(r'-ik'a^2)^2}\, dr'^2.$$

Defining $u = r'-ik'a^2$, $r' = u+ik'a^2$, $dr' = du$, $dr'^2 = 2(u+ik'a^2)du$. With this substitution

$$f_B(k') = -\frac{mV_0}{2\hbar^2 k'}\, e^{-\frac{k'^2a^2}{2}}\int_{-\infty}^{+\infty} e^{-\frac{u^2}{2a^2}}\, 2k'a^2 du =$$

$$= -\frac{mV_0 a^2\sqrt{\pi}}{\hbar^2}\,\sqrt{2}\, a e^{\frac{-k'^2a^2}{2}}$$

i.e.

$$f_B(k') = -\frac{\sqrt{2\pi}\, mV_0 a^3\, e^{-k'^2a^2/2}}{\hbar^2}.$$

The differential scattering cross section in this approximation is

$$\frac{d\sigma}{d\Omega} = \frac{2\pi m^2 V_0^2 a^6}{\hbar^4}\, e^{-k'^2a^2},$$

and the total cross section

$$\sigma = \frac{4\pi^2 m^2 V_0^2 a^6}{\hbar^4}\int_{\theta=0}^{\theta=\pi} d\cos\theta\, e^{-4a^2k^2\sin^2\theta/2} =$$

$$= -\frac{4\pi^2 m^2 V_0^2 a^6}{\hbar^4}\, e^{-2k^2a^2}\int_{\theta=0}^{\theta=\pi} d\cos\theta\, e^{2a^2k^2\cos\theta}.$$

$$\sigma = \frac{2\pi^2 m^2 V_0^2 a^4}{k^2 \hbar^4}\left(1 - e^{-4k^2a^2}\right) \underset{k\to\infty}{\to} \frac{2\pi^2 m^2 V_0^2 a^4}{k^2 \hbar^4},$$

a result independent of the sign of V_0. This problem may also be solved using Cartesian coordinates.

PROBLEM 6.7. Find the reflection time for a particle incident with energy E on a step barrier $V = 0$ $x < 0$, $V = V_0$ $x > 0$ $(V_0 > 0)$, $E < V_0$.

$$\text{For } x < 0 \quad \psi = Ae^{ikx} + Be^{-ikx} ;$$

$$\text{for } x > 0 \quad \psi = Ce^{-\eta x} ,$$

where

$$k = \sqrt{\frac{2mE}{\hbar^2}}, \qquad \eta = \sqrt{\frac{2m(V_0-E)}{\hbar^2}} = \sqrt{\frac{2mV_0}{\hbar^2} - k^2} = \eta(k).$$

Matching boundary conditions at $x = 0$ one obtains

$$B = A\,\frac{1 - \frac{i\eta}{k}}{1 + \frac{i\eta}{k}} .$$

Constructing wavepackets:

$$\Psi_{in}(x,t) = \int A(k')\, e^{i(k'x - E(k')t/\hbar)}\, dk'$$

$$\Psi_{ref}(x,t) = \int B(k')\, e^{-i(k'x+E(k')t/\hbar)}\, dk'$$

$$= \int A(k')\, \frac{k'-i\eta(k')}{k'+i\eta(k')}\, e^{-i(k'x+E(k')t/\hbar)}\, dk'$$

$$= \int A(k')e^{-i(k'x+E(k')t/\hbar+\delta(k'))}\, dk'$$

where

$$E(k) = \frac{\hbar^2k^2}{2m}, \quad \delta(k) = \tan^{-1}\frac{2\eta(k)k}{k^2-\eta^2(k)} = 2\tan^{-1}\frac{\eta(k)}{k} .$$

Using stationary phase arguments (cf. Equations (4.14)) one obtains

$$\Psi_{in}(x,t) = e^{i\phi}\,\Psi_{in}(x - v_g t,\ 0)$$

$$\Psi_{ref}(x,t) = e^{i\theta}\Psi_{ref}\left(x + v_g t + \left.\frac{d\delta(k')}{dk'}\right|_{k'=k},\ 0\right).$$

If at t = 0 the particle reaches x = 0, it leaves

$$x = 0 \text{ at } \tau = -\frac{1}{V_g}\left.\frac{d\delta(k')}{dk'}\right|_{k'=k} = \frac{2m}{\hbar k\eta}$$

since

$$V_g = \frac{\hbar k}{m}\ .$$

If $\eta \to \infty$ i.e. $V_0 \to \infty$, $\tau \to 0$. In this limit B = -A. τ is a minimum if $k = \eta = \sqrt{(mV_0/\hbar^2)}$. Moreover δ is discontinuous at $k = \sqrt{(2mV_0/\hbar^2)}$ and its derivative is discontinuous at k = 0.

PROBLEM 6.8. Consider a particle scattered off two identical delta function potentials namely

$$V(x) = -|V_0|\left\{\delta(x-a) + \delta(x+a)\right\}$$

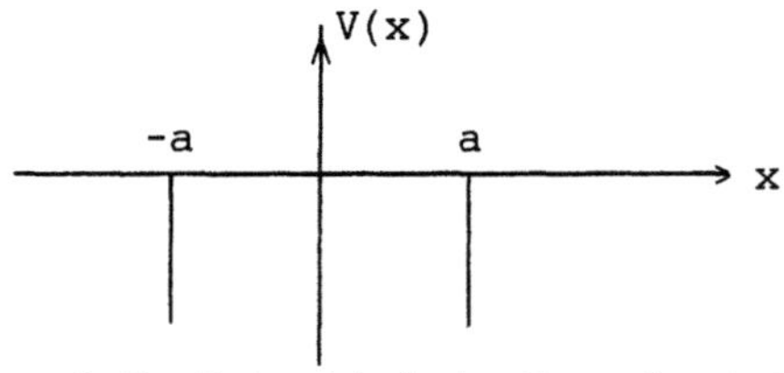

Figure 6.7. Potential in Example 6.8.

Find the scattering amplitude and its poles.

For convenience one may choose

$$\Psi(x) = \begin{cases} Ae^{ik(x+a)} + Be^{-ik(x+a)} & -\infty < x < -a \\ Ce^{ik(x+a)} + De^{-ik(x+a)} & -a < x < a \\ Fe^{ik(x-a)} & x > a\ . \end{cases}$$

The wavefunction is continuous at x = ± a and its derivative satisfies equations analogous to Equation (6.10), namely:

$$\lim_{\delta\to 0}\left\{\left.\frac{d\Psi}{dx}\right|_{x=\pm a+\delta} - \left.\frac{d\Psi}{dx}\right|_{x=\pm a-\delta}\right\} = -\frac{2m|V_0|}{\hbar^2}\,\Psi(x=\pm a).$$

Hence, defining

$$\varepsilon \equiv \frac{m|V_0|}{\hbar^2 k}$$

one obtains that:

$$A+B = C+D$$

$$A(1+2i\varepsilon) + B(-1+2i\varepsilon) = C - D$$

$$Ce^{2ika} + De^{-2ika} = F$$

$$Ce^{2ika} - De^{-2ika} = F(1-2i\varepsilon)$$

which imply

$$A = (1-i\varepsilon)\, C - i\varepsilon\, D$$

$$C = e^{-2ika}(1-i\varepsilon)\, F$$

$$D = i\varepsilon\, e^{2ika}\, F.$$

Solving for F/A one obtains

$$\frac{F}{A} = \left\{(1-i\varepsilon)^2\, e^{-2ika} + \varepsilon^2\, e^{2ika}\right\}^{-1}$$

$$= \left\{(1-i2\varepsilon)\cos 2ka + i(2\varepsilon^2 - 1 + 2i\varepsilon)\sin 2ka\right\}^{-1}.$$

The scattering amplitude is thus

$$\frac{Fe^{-ika}}{Ae^{ika}} = \frac{F}{A}\, e^{-2ika} = e^{-2ika}\,[\cos 2ka - 2\varepsilon \sin 2ka + i((2\varepsilon^2-1)\sin 2ka - 2\varepsilon\cos 2ka)]^{-1} \tag{6.36}$$

$$= \frac{e^{-i(2ka+\delta)}}{\sqrt{(\cos 2ka - 2\varepsilon\sin 2ka)^2 + ((2\varepsilon^2-1)\sin 2ka - 2\varepsilon\cos 2ka)^2}},$$

with

$$\delta = \tan^{-1}\left\{(2\varepsilon^2-1)\tan 2ka - 2\varepsilon\right\}\left\{1 - 2\varepsilon\tan 2\,ka\right\}^{-1}.$$

If $a \to 0$ Equation (6.36) reduces to Equation (6.13) as it must (but with V_0 replaced by $2V_0$), namely $T = 1/(1-i2\varepsilon)$, $\delta = -\tan^{-1}2\varepsilon$.

The poles of Equation (6.36) occur when

$$-i\tan 2ka = \frac{1-i2\varepsilon}{2\varepsilon^2-1+2i\varepsilon} \tag{6.37}$$

This equation is satisfied if $E < 0$ since then

$$k = \sqrt{\frac{2mE}{\hbar^2}} = i\sqrt{\frac{2m|E|}{\hbar^2}} \equiv i\kappa$$

$$\varepsilon = \frac{m|V_0|}{\hbar^2 i\kappa} \equiv -i\epsilon.$$

Equation (6.37) then reduces to the equation for the allowed bound states

$$\tanh 2\kappa a = \frac{2\epsilon-1}{1-2\epsilon+2\epsilon^2},$$

which is satisfied when either

$$\tanh \kappa a = 2\epsilon - 1 \qquad (6.38a)$$

or

$$\tanh \kappa a = (2\epsilon-1)^{-1} \qquad (6.38b)$$

Equation (6.38a) corresponds to even parity bound states while (6.38b) to odd parity bound states.

If $a \to 0$ Equation (6.38b) has no solution while Equation (6.38a) has only one solution namely

$$2\epsilon - 1 = 0 \quad \text{i.e.} \quad \epsilon = \frac{1}{2}, \quad \text{that is } |E| = \frac{m(2|V_0|)^2}{2\hbar^2}$$

which is just Equation (6.15) with V_0 replaced by $2V_0$, while if $a \to \infty$ both Equation (6.38a) and Equation (6.38b) reduce to $\epsilon = 1$ which is just Equation (6.15). These results are as expected since in the former case the delta function potentials coalesce while the latter they essentially uncouple.

If k becomes large ε tends to zero and δ to -2ka. This means if the incident wavepacket is $\Psi_{inc}(x+a-v_g t, 0)$ i.e. $x = -a$ at $t = 0$, Ψ_{trans} is $\Psi_{trans}(x+a-v_g t, 0)$ i.e. $x = a$ at $t = 2a/v_g$ which as expected is the classical transit time for crossing these two potentials.

CHAPTER 7

Heisenberg Representation

Starting with the expression for the expectation value of an operator O_S in the Schrödinger representation (at some time t)

$$<O>_t = \int \Psi^*(x, t)\, O_S\, \Psi(x, t)\, dx \tag{7.1}$$

and the fact that generally (Equation (4.12))

$$\Psi(x, t) = \int G(x, x', t)\Psi(x', 0)\, dx' \tag{7.2}$$

where for a free particle

$$G(x, x', t) = \int \frac{1}{\sqrt{2\pi}}\, e^{-ip'x'/\hbar}\, \frac{1}{\sqrt{2\pi}}\, e^{ip'x/\hbar}\, e^{-iE(p')t/\hbar}\, \frac{dp'}{\hbar} \tag{7.3a}$$

and by analogy for a particle in a (time independent) potential

$$G(x, x', t) = \sum_{n=0}^{\infty} \phi_n^*(x')\, \phi_n(x)\, e^{-iE_n t/\hbar} \tag{7.3b}$$

one can go from the Schrödinger to the Heisenberg representation.

The procedure involves first noting that Equation (7.3a) can be written

$$G(x, x', t) = \int \frac{1}{\sqrt{2\pi}}\, e^{-ip'x'/\hbar}\, e^{-iH(x)t/\hbar}\, \frac{1}{\sqrt{2\pi}}\, e^{ip'x/\hbar}\, dp'/\hbar$$

$$= e^{-iH(x)t/\hbar} \int \frac{1}{\sqrt{2\pi}}\, e^{-ip'x'/\hbar}\, \frac{1}{\sqrt{2\pi}}\, e^{ip'x/\hbar}\, \frac{dp'}{\hbar} = e^{-iH(x)t/\hbar}\, \delta(x-x')$$

(cf. Equation (2.19)).

Similarly Equation (7.3b) can be written

$$G(x, x', t) = \sum_n \phi_n^*(x') \, e^{-iH(x)t/\hbar} \, \phi_n(x) = e^{-iH(x)t/\hbar} \, \delta(x-x'),$$

(cf. Equation (2.6)), i.e. generally

$$G(x, x', t) = e^{-iH(x)t/\hbar} \, \delta(x-x'). \tag{7.4}$$

Substituting the result (7.4) back into expression (7.2) yields:

$$\boxed{\Psi(x, t) = e^{-iH(x)t/\hbar} \, \Psi(x, 0)} \tag{7.5}$$

Thus Equation (7.1) can be rewritten

$$\langle O \rangle_t = \int \Psi^*(x, 0) \, e^{iH(x)t/\hbar} \, O_S e^{-iH(x)t/\hbar} \, \Psi(x, 0) \, dx, \tag{7.6}$$

i.e.

$$\langle O \rangle_t = \int \Psi^*(x, 0) \, O_H \, \Psi(x, 0) \, dx \tag{7.7}$$

where

$$\boxed{O_H = e^{iH(x)t/\hbar} \, O_S \, e^{-iH(x)t/\hbar}} \tag{7.8}$$

and the subscripts S and H stand for the operator in the Schrödinger and Heisenberg representations respectively.

In Equation (7.7) one thus has simpler wavefunctions than in Equation (7.1) i.e. only $\Psi(x, 0)$, independent of the time, but more complicated operators O_H rather than O_S.

The operator O_H can be seen to satisfy the differential equation:

$$\boxed{\frac{dO_H}{dt} = \frac{i}{\hbar} [H_H, O_H] + \frac{\partial O_H}{\partial t},} \tag{7.9}$$

since

$$\frac{dO_H}{dt} = \frac{i}{\hbar} e^{iHt/\hbar} (HO_S - O_S H) e^{-iHt/\hbar} + e^{iHt/\hbar} \frac{\partial O_S}{\partial t} e^{-iHt/\hbar}$$

and where

$$\frac{\partial O_H}{\partial t} \equiv e^{iHt/\hbar} \frac{\partial O_S}{\partial t} e^{-iHt/\hbar} .$$

To obtain Equation (7.9) one also must insert $e^{-iHt/\hbar} e^{iHt/\hbar} = 1$ between H and O_S. One assumes here that $H \neq H(t)$ in which case

$$H_H = e^{iH_S(x)t/\hbar} H_S e^{-iH_S(x)t/\hbar} = H_S .$$

If $O = x$ or p and $H = p^2/2m + V(x)$,

then

$$\frac{dx_H}{dt} = \frac{i}{\hbar} [H_H, x_H] = \frac{p_H}{m} , \qquad (7.10)$$

$$\frac{dp_H}{dt} = \frac{i}{\hbar} [H_H, p_H] = - \frac{\partial V_H}{\partial x} = F_H , \qquad (7.11)$$

and combining Equation (7.10) and (7.11) one obtains

$$m \frac{d^2 x_H}{dt^2} = - \frac{\partial V_H}{\partial x} = F_H . \qquad (7.12)$$

The standard Heisenberg representation results (7.10) - (7.12) look exactly like the corresponding classical expressions for velocity and force. In fact Equation (7.12) looks just like Newton's 2nd law. This analogy has formal merit.

A word of caution is however in order here. p_H and x_H are quantum mechanical operators and are more complicated than their classical analogues. Thus

$$x_H p_H \neq p_H x_H, \qquad \text{but since } [x_H, p_H] = - \frac{\hbar}{i} \qquad (7.13)$$

$$x_H p_H = p_H x_H - \frac{\hbar}{i} \qquad (7.14)$$

EXAMPLE 7.1. Show that generally

$$\frac{d(O^2)_H}{dt} = \frac{d(O_H)^2}{dt} \neq 2O_H \frac{dO_H}{dt}$$

but rather that

$$\frac{d(O^2)_H}{dt} = \frac{d(O_H)^2}{dt} = 2O_H \frac{dO_H}{dt} + \left[\frac{dO_H}{dt}, O_H\right] .$$

From Equation (7.9)

$$\frac{d(O_H)^2}{dt} = \frac{i}{\hbar}[H_H, (O_H)^2] \text{ if } (O_S)^2$$

does not depend on t explicitly,
where

$$(O^2)_H = e^{iHt/\hbar} O_S^2 e^{-iHt/\hbar} = e^{iHt/\hbar} O_S e^{-iHt/\hbar} e^{iHt/\hbar} O_S e^{-iHt/\hbar}$$
$$= (O_H)^2,$$

and

$$[H_H, (O_H)^2] = [H_H, O_H] O_H + O_H[H_H, O_H].$$

Hence

$$\frac{dO_H^2}{dt} = \frac{i}{\hbar}[H_H, O_H]O_H + O_H \frac{i}{\hbar}[H_H, O_H]$$
$$= \frac{dO_H}{dt} O_H + O_H \frac{dO_H}{dt} = 2O_H \frac{dO_H}{dt} + \left[\frac{dO_H}{dt}, O_H\right]$$

i.e.

$$\frac{d(O_H)^2}{dt} = \frac{d(O^2)_H}{dt} = 2O_H \frac{dO_H}{dt} + \left[\frac{dO_H}{dt}, O_H\right]. \tag{7.15}$$

EXAMPLE 7.2. Show

$$\frac{d(x^2)_H}{dt} = \frac{d(x_H)^2}{dt} = 2x_H \frac{dx_H}{dt} + \frac{\hbar}{mi} .$$

From Equation (7.15),

$$\frac{d(x^2)_H}{dt} = \frac{d(x_H)^2}{dt} = 2x_H \frac{dx_H}{dt} + \left[\frac{dx_H}{dt}, x_H\right].$$

But from Equation (7.10)

$$\frac{dx_H}{dt} = \frac{p_H}{m} ,$$

thus

$$\frac{dx_H^2}{dt} = 2x_H \frac{dx_H}{dt} + \frac{1}{m}\left[p_H, x_H\right] = 2x_H \frac{dx_H}{dt} + \frac{\hbar}{mi}, \tag{7.16}$$

where one has also used the result (7.13).
One notes in passing that one readily gets Ehrenfest's expressions from the preceding since

$$\frac{d\langle O\rangle_t}{dt} = \int \Psi^*(x, 0) \frac{dO_H}{dt} \Psi(x, 0)\, dx = \langle \frac{dO_H}{dt} \rangle$$

$$= \frac{i}{\hbar} \langle [H_H, O_H] \rangle + \langle \frac{\partial O_H}{\partial t} \rangle . \tag{7.17}$$

EXAMPLE 7.3. Evaluate $d/dt \langle \frac{1}{2} m\omega^2 x_H^2 \rangle$ for a particle in the ground state of a simple harmonic oscillator potential $V(x) = \frac{1}{2} m\omega^2 x^2$.
From Equation (7.16)

$$\frac{dx_H^2}{dt} = 2x_H \frac{dx_H}{dt} + \frac{\hbar}{mi},$$

therefore

$$\langle \frac{dx_H^2}{dt} \rangle = \frac{2}{m} \int \phi_0^*(x)\, x_H p_H\, \phi_0(x)\, dx + \frac{\hbar}{mi} .$$

But

$$\int \phi_0^*(x)\, x_H p_H \phi_0(x)\, dx = \int \phi_0^*(x)\, x \frac{\hbar}{i} \frac{\partial}{\partial x} \phi_0(x)\, dx =$$

$$- \frac{\hbar}{i} \int \phi_0^*(x)\, x \frac{m\omega x}{\hbar} \phi_0(x)\, dx = - \frac{2}{i\omega} \int \phi_0^*(x) \frac{m\omega^2 x^2}{2} \phi_0(x)\, dx = -\frac{\hbar}{2i}$$

where one has used the fact that $\phi_0(x) = Ne^{-m\omega x^2/2\hbar}$.
Hence

$$\langle \frac{dx_H^2}{dt} \rangle = - \frac{\hbar}{im} + \frac{\hbar}{im} = 0,$$

and the expectation value of the kinetic energy of the ground (and by a similar calculation of any) state of a simple harmonic oscillator is independent of t.

EXAMPLE 7.4. Generalize the result of Example 7.3 for any state of the simple harmonic oscillator.

Defining

$$a_H^{\pm} \equiv \frac{1}{\sqrt{2m}} \left\{ p_H \pm im\omega x_H \right\} ,$$

one can quickly show, using Equation (7.13) that $[a_H^+, a_H^-] = -\hbar\omega$, and that

$$x_H p_H = \frac{1}{2\omega i} \left\{ (a_H^+)^2 - (a_H^-)^2 + [a_H^+, a_H^-] \right\}$$

$$= \frac{1}{2\omega i} \left\{ (a_H^+)^2 - (a_H^-)^2 - \hbar\omega \right\} .$$

Thus

$$< \frac{dX_H^2}{dt} > = \frac{2}{m} \int \phi_n^* (x)\, x_H p_H\, \phi_n(x)\, dx + \frac{\hbar}{mi}$$

$$= \frac{1}{im\omega} \int \phi_n^*(x) \left\{ (a_H^+)^2 - (a_H^-)^2 - \hbar\omega \right\} \phi_n(x)\, dx + \frac{\hbar}{mi}$$

$$= - \frac{\hbar}{mi} + \frac{\hbar}{mi} = 0,$$

since $(a_H^{\pm})^2$ operating on $\phi_n(x)$ produce states orthogonal to $\phi_n(x)$. For non-diagonal matrix elements (i.e. $n \neq m$) $\frac{d}{dt} < n|\frac{1}{2}m\omega^2 x_H^2|m> \neq 0$.

EXAMPLE 7.5. Show

$$\frac{dT_H}{dt} = \frac{p_H F_H}{m} + \frac{1}{2m} [F_H, p_H], \tag{7.18}$$

where T_H is the kinetic energy operator in the Heisenberg representation.

From Equation (7.15)

$$\frac{d(p_H)^2}{dt} = \frac{d(p^2)_H}{dt} = 2p_H \frac{dp_H}{dt} + \left[\frac{dp_H}{dt} , p_H \right] .$$

But from Equation (7.11) $\frac{dp_H}{dt} = - \frac{\partial V_H}{\partial x} = F_H.$

$$\therefore \frac{dp_H^2}{dt} = 2p_H F_H + [F_H, p_H]$$

or

$$\frac{dT_H}{dt} = \frac{p_H F_H}{m} + \frac{1}{2m}[F_H, p_H].$$

Thus for example if $V_H = \frac{1}{2} m\omega^2 x_H^2$

$$\frac{dp_H}{dt} = - \frac{\partial V_H}{\partial x} = F_H = - m\omega^2 x_H$$

and

$$\frac{dT_H}{dt} = - \omega^2 p_H x_H + \frac{\hbar\omega^2}{2i} \quad .$$

CHAPTER 8

Two and Three Versus One-Dimensional Problems

The one-dimensional Schrödinger equation for a particle in a potential $V_1(x)$ is

$$\left\{-\frac{\hbar^2}{2m}\frac{d^2}{dx^2}+V_1(x)\right\}\psi_n(x) = E_n\psi_n(x) , \qquad (8.1)$$

where

$$\int_{-\infty}^{+\infty} |\psi_n(x)|^2\, dx = 1.$$

The radial equation for a particle in a three dimensional radial potential $V_3(r)$ is

$$\left\{-\frac{\hbar^2}{2m}\frac{d^2}{dr^2}+\left(V_3(r)+\frac{\hbar^2\ell(\ell+1)}{2mr^2}\right)\right\} u_{n\ell}(r) = E_{n\ell}u_{n\ell}(r), \qquad (8.2)$$

where

$$\ell = 0, 1, 2, \ldots, \qquad \psi_{n\ell m}(r, \theta, \phi) = \frac{u_{n\ell}(r)}{r} Y^{\ell}_{m}(\theta, \phi) -\ell\le m\le \ell,$$

and

$$\int_0^{\infty} u^2_{n\ell}(r)\, dr = 1.$$

If $\ell = 0$ and $V_1(x) = V_3(x)$ Equations (8.1) and (8.2) are identical. However, $\psi_n(x)$ in Equation (8.1) extends from $-\infty \le x \le \infty$ while $u_{n\ell}(r)$ must be zero at $r = 0$ in order that $u_{n\ell}(r)/r$ be finite at the origin. In addition $u_{n\ell}(r)$ extends only from $r = 0$ to $r = \infty$. A consequence of the above is that some solutions acceptable for Equation (8.1) are not acceptable for Equation (8.2). Radial solutions of Equation (8.2) with potential $V_3(r)$ are identical with solutions of the one dimensional problem (8.1) if:

$$V_1(x) = \infty \quad x < 0$$

$$V_1(x) = V_3(x) + \frac{\hbar^2 \ell(\ell+1)}{2mx^2} \quad x > 0. \tag{8.3}$$

If

$$V_1(x) = V_3(x) + \frac{\hbar^2 \ell(\ell+1)}{2mr^2}$$

and moreover $V_1(x) = V_1(-x)$ then the odd parity solutions (i.e. those solutions which vanish at x = 0) of $\psi_n(x)$ are identical to $u_{n'\ell}(x)/\sqrt{2}$.

EXAMPLE 8.1. Consider the potential $V_1(x) = (\hbar^2/2mb^4)x^2$. For this potential the solutions of Equation (8.1) for the energy E_n are $E_n = (n+\frac{1}{2})\,\hbar^2/mb^2$. The lowest energy is $\hbar^2/2mb^2$ and the ground-state wavefunction $(1/\sqrt{\pi}b)^{\frac{1}{2}}\,e^{-x^2/2b^2} = \psi_0(x)$.

Consider now the potential $V_3(r) = \hbar^2/2mb^4\; r^2$ and the case $\ell = 0$. The minimum energy for a particle in this potential is $3\hbar^2/2mb^2$. Examine why this is so.

$u_{00}(r)$ cannot be $\psi_0(r)$ since $\psi_0(x)$ does not vanish at x = 0. The second energy eigenvalue for the one-dimensional problem is $E_1 = 3\hbar^2/\,2mb^2$, with eigenfunction

$$\psi_1(x) = \left(\frac{2}{b\sqrt{\pi}}\right)^{\frac{1}{2}} \left(\frac{x}{b}\right) e^{-x^2/2b^2}.$$

This is acceptable as the lowest eigenfunction of Equation (8.2) since $\psi_1(0)=0$. The normalization must be modified however since

$$\int_{-\infty}^{+\infty} \psi_1^2(x)\, dx = 1$$

while

$$\int_0^{\infty} u_{00}^2(r)\, dr = 1.$$

Thus

$$u_{00}(r) = \frac{2e^{-r^2/2b^2}}{(b\sqrt{\pi})^{\frac{1}{2}}}\left(\frac{r}{b}\right) .$$

Similarly all odd solutions $\psi_3(x)$, $\psi_5(x)$ are acceptable (see Example 10.3) solutions since for these solutions $\psi(0) = 0$. Thus $u_{10}(r) = \sqrt{2}\ \psi_3(r)$, $u_{20}(r) = \sqrt{2}\ \psi_5(r)$ etc. and

$$E_{\frac{n-1}{2},\ 0} = (n+\tfrac{1}{2})\frac{\hbar^2}{mb^2} \qquad n = 1, 3, 5 \ldots$$

i.e.

$$E_{p,\ 0} = (2p + 3/2)\hbar^2/mb^2, \qquad p = 0, 1, 2\ldots$$

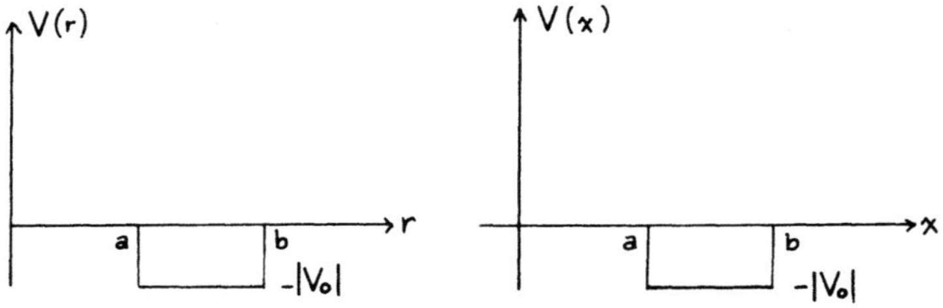

Figure 8.1. Diagram in Example 8.2.

EXAMPLE 8.2. Consider a particle in the three-dimensional well $V_3(r) = 0\ \ 0 < r < a$; $V_3(r) = -|V_0|$, $a < r < b$, $V_3(r) = 0\ \ r > b$. Compare this and the analogous one-dimensional system

$$V_1(x) = -|V_0|, \quad a < x < b \quad V_1(x) = 0 \quad x < a, \quad x > b.$$

The allowed $\ell = 0$ energy levels are easy to obtain. In detail the solutions of Equation (8.2) which satisfy the boundary conditions at the origin ($u_{n0}(0) = 0$) are

$$0 < r < a \quad u_{n0}(r) = A \sinh kr \qquad k = \sqrt{\frac{2m|E|}{\hbar^2}} ,$$

$$K = \sqrt{\frac{2m(|V_0|-|E|)}{\hbar^2}}$$

$$a < r < b \qquad u_{n0}(r) = B \sin(Kr+\delta)$$

$$b < r \qquad u_{n0}(r) = Ce^{-kr} .$$

Matching the wavefunctions and their derivatives at a and b yields

$$\frac{\tanh ka}{k} = \frac{\tan(Ka+\delta)}{K}$$

i.e.

$$\tan(Ka+\delta) = \frac{K}{k}\tanh ka \tag{8.4}$$

and

$$\frac{\tan(Kb+\delta)}{K} = -\frac{1}{k}$$

i.e.

$$\tan(Kb+\delta) = -\frac{K}{k}. \tag{8.5}$$

One can rewrite Equation (8.5) as

$$\tan\,(Ka+\delta+K(b-a)) = -\frac{K}{k} = \frac{\tan(Ka+\delta)+\tan K(b-a)}{1-\tan(Ka+\delta)\tan K(b-a)} = \frac{K/k\tanh ka+\tan K(b-a)}{1-\frac{K}{k}\tanh ka\tan K(b-a)}\ .$$

Thus the energy levels for this system are obtained by solving the transcendental equation

$$-\frac{K}{k} = \frac{K/k\tanh ka+\tan K(b-a)}{1-\frac{K}{k}\tanh ka\tan K(b-a)}\ . \tag{8.6}$$

If $a = 0$ Equation (8.6) reduces to $-K/k = \tan K b$.
If $a \to \infty$, $b \to \infty$, $c \equiv b-a$ Equation (8.6) reduces to

$$-\frac{K}{k} = \frac{K/k + \tan Kc}{1-K/k\tan Kc}\ .$$

These results can be compared with the energy levels for a particle in the one dimensional well given in this example. The solutions of Equation (8.1) for this potential are

$$\psi(x) = Ae^{k|x|} \qquad x < a$$

$$\psi(x) = B\sin\,(Kx+\delta) \qquad a < x < b$$

$$\psi(x) = Ce^{-kx} \qquad x > b.$$

Matching boundary conditions one obtains

$$\frac{1}{k} = \frac{\tan(Ka+\delta)}{K} \tag{8.7}$$

$$\frac{\tan(Kb+\delta)}{K} = -\frac{1}{k}. \tag{8.8}$$

One can rewrite Equation (8.8) as

$$\tan((Ka+\delta)+K(b-a)) = -\frac{K}{k} = \frac{\tan(Ka+\delta)+\tan K(b-a)}{1-\tan(Ka+\delta)\tan K(b-a)} = \frac{K/k + \tan K(b-a)}{1 - \frac{K}{k}\tan K(b-a)} .$$

Thus the energy levels for this system are obtained by solving the transcendental equation

$$-\frac{K}{k} = \frac{K/k + \tan K(b-a)}{1-K/k \tan K(b-a)} . \tag{8.9}$$

This is not equal to Equation (8.6) except in the limit

$a \to \infty$ $\quad b - a \equiv c \quad$ where c is
$b \to \infty$

a constant, in which case the different boundary condition at the origin is unimportant.

The two-dimensional Schrödinger equation for a particle in a potential $V_2(\rho)$ can be written:

$$\left\{-\frac{\hbar^2}{2m}\frac{d^2}{d\rho^2} + \left(V_2(\rho) + \frac{\hbar^2(m^2-\frac{1}{4})}{2M\rho^2}\right)\right\} w_{nm}(\rho) = E_{nm}w_{nm}(\rho) \tag{8.10}$$

where

$$\psi_{nm}(\rho, \phi) = \frac{w_{nm}(\rho)}{\rho^{\frac{1}{2}}}\ \frac{e^{\pm im\phi}}{\sqrt{2\pi}}, \qquad m = 0, \pm 1, \pm 2 ...$$

and

$$\int_0^\infty w_{nm}^2(\rho)\, d\rho = 1.$$

Here $w_{nm}(\rho)$ must be zero at

$$\rho = 0 \text{ so } \frac{w_{nm}(\rho)}{\rho^{\frac{1}{2}}}$$

will be finite at the origin.

Comparing Equations (8.2) and (8.10) and noting that $m^2 - \frac{1}{4} = (m - \frac{1}{2})(m - \frac{1}{2} + 1)$ indicates that if one has a solution $u_{n\ell}(r)$ of Equation (8.2), then $w_{nm}(\rho)$ is $u_{nm-\frac{1}{2}}(\rho)$. One notes both $u(0)$ and $w(0)$ must be zero as too $U(\infty)$, and $w(\infty)$, i.e. $u(r)$ and $w(\rho)$ have identical boundary conditions.

EXAMPLE 8.3. Suppose one has a particle of mass M in a two-dimensional Coulomb potential

$$V_2(\rho) = -\frac{Ze^2}{4\pi\varepsilon_0\rho} .$$

Find the corresponding energies and wavefunctions. Since, as standard texts report, for the three dimensional potential $V_3(r) = -\frac{Ze^2}{4\pi\varepsilon_0 r}$

$$u_{n\ell}(r) = \frac{e^{-Kr}(2Kr)^{\ell+1}}{(2\ell+1)!} \left[\frac{K(n+\ell)!}{n(n-\ell-1)!}\right]^{\frac{1}{2}} {}_1F_1(-n+\ell+1;\ 2\ell+2;\ 2Kr), \tag{8.11}$$

and

$$E_{n\ell} = -\frac{Z^2\alpha^2Mc^2}{2n^2},\ K = \frac{ZMc\alpha}{\hbar n},\ \alpha = \frac{e^2}{4\pi\varepsilon_0\hbar c} \quad \begin{matrix} n = 1, 2\ldots \\ \ell = 0, 1, \ldots n-1 \end{matrix},$$

one can immediately write

$$w_{nm}(\rho) = e^{-K\rho}\frac{(2K\rho)^{m+\frac{1}{2}}}{(2m)!} \left[\frac{K(n+m-\frac{1}{2})!}{n(n-m-\frac{1}{2})!}\right]^{\frac{1}{2}} {}_1F_1(-n+m+\tfrac{1}{2};\ 2m+1;\ 2K\rho). \tag{8.12}$$

$$E_{nm} = -\frac{Z^2\alpha^2Mc^2}{2n^2}, \quad \begin{matrix} n = \frac{1}{2}, \frac{3}{2} \ldots \\ m = 0, 1, 2, \ldots n - \frac{1}{2} \end{matrix}.$$

Thus

$$E_{nm} = -2Z^2\alpha^2Mc^2,\ -\frac{2Z^2\alpha^2Mc^2}{9},\ \ldots$$

EXAMPLE 8.4. Suppose one has a particle of mass M in a two-dimensional oscillator potential $V_2(\rho) = 1/2M\omega^2\rho^2$. Find the corresponding energies and wavefunctions.

Since as standard texts report for the three dimensional potential $V_3(r) = \frac{1}{2}M\omega^2r^2$,

$$u_{n\ell}(r) = \left\{\frac{2\Gamma(n+\ell+\frac{3}{2})}{b^3n!}\right\}^{\frac{1}{2}} \frac{r^{\ell+1}e^{-r^2/2b^2}}{b^\ell\,\Gamma(\ell+\frac{3}{2})}\ {}_1F_1(-n,\ \ell+\tfrac{3}{2},\ r^2/b^2), \tag{8.13}$$

$$E_{n\ell} = (2n + \ell + \tfrac{3}{2})\hbar\omega,\ b = \sqrt{\frac{\hbar}{M\omega}} \quad \begin{matrix} n = 0, 1, 2\ldots \\ \ell = 0, 1\ldots \end{matrix}$$

one can immediately write

$$w_{nm}(\rho) = \left\{\frac{2(n+m)!}{b^2n!}\right\}^{\frac{1}{2}} \frac{\rho^{m+\frac{1}{2}}e^{-\rho^2/2b^2}}{b^m\,m!}\ {}_1F_1\left(-n;\ m+1;\ \frac{\rho^2}{b^2}\right). \tag{8.14}$$

$$E_{nm} = (2n + m + 1)\hbar\omega, \quad b = \sqrt{\frac{\hbar}{M\omega}} \quad \begin{matrix} n = 0, 1\ldots \\ m = 0, 1\ldots \end{matrix}$$

EXAMPLE 8.5. Suppose one has a particle of mass M in a two-dimensional potential $V = 0$ $0 < \rho < a$, $= \infty$ $\rho > a$. Solve this problem.

To find the eigenvalues and eigenfunctions of this system one notes that for the analogous three dimensional case the standard result involves spherical Bessel and spherical harmonic functions:

$$\psi_{n\ell m}(r, \theta, \phi) = A_\ell j_\ell(kr)\, Y_m^\ell(\theta, \phi) \qquad 0 < r < a$$
$$= 0 \qquad r > a, \quad k = \sqrt{\frac{2ME}{\hbar^2}}$$

with allowed energies corresponding to values of k such that $j_\ell(ka) = 0$.

For the two-dimensional case therefore

$$\Psi_{nm}(\rho, \phi) \propto \rho^{\frac{1}{2}}\, j_{n-\frac{1}{2}}(k\rho)\, e^{\pm im\phi}$$

with allowed energies when k is such that $j_{\ell-\frac{1}{2}}(ka) = 0$. One notes that

$$j_\ell(z) = \sqrt{\frac{\pi}{2z}}\, J_{\ell+\frac{1}{2}}(z) .$$

Hence these results may be written $\psi_{nm}(\rho, \phi) = C_n J_n(k\rho) e^{\pm im\phi}$ with allowed energies when the cylindrical Bessel function $J_n(ka) = 0$.

EXAMPLE 8.6. Treat the finite square well system in two dimensions by analogy with the standard three dimensional results:

$$\psi_{int}(r, \theta, \phi) = A_\ell j_\ell(Kr)\, Y_m^\ell(\theta, \phi), \qquad 0 < r < a,$$

$$\psi_{ext}(r, \theta, \phi) = B_\ell h_\ell^{(1)}(iKr) Y_m^\ell(\theta, \phi) \qquad r > a,$$

$$K = \sqrt{\frac{2M}{\hbar^2}(|V_0| - |E|)}\,, \quad k = \sqrt{\frac{2M|E|}{\hbar^2}} .$$

By analogy:

$$\psi_{nm\,int.}(\rho, \phi) \propto \rho^{\frac{1}{2}}\, j_{n-\frac{1}{2}}(K\rho)\, e^{\pm im\phi} \qquad 0 < \rho < a,$$

$$\psi_{nm\,ext.}(\rho, \phi) \propto \rho^{\frac{1}{2}} h_{n-\frac{1}{2}}^{(1)}(ik\rho)\, e^{\pm im\phi} \qquad a < \rho < \infty.$$

But

$$j_{\ell-\frac{1}{2}}(z) = \sqrt{\frac{\pi}{2z}}\, J_\ell(z), \quad h_{\ell-\frac{1}{2}}^{(1)}(z) = \sqrt{\frac{\pi}{2z}}\, H_\ell^{(1)}(z).$$

Therefore,

$$\psi_{nm\,int.}(\rho, \phi) = D_n\, J_n(K\rho)\, e^{\pm im\phi} \qquad 0 < \rho < a,$$

$$\psi_{nm\,\text{ext.}}(\rho,\ \phi) = F_n H_n^{(1)}(ik\rho)\ e^{\pm im\phi} \qquad a < \rho < \infty.$$

Matching the wavefunctions and derivatives at $\rho = a$ gives one the appropriate quantization condition for the energy.

CHAPTER 9

'Kramer' Type Expressions, The Virial Theorem and Generalizations

Consider a particle moving in a central potential $V(r) = Ar^p$. The (radial) differential equation for $u_{n\ell}(r) = r\, R_{n\ell}(r)$, (where the complete wave-function $\psi(r, \theta, \phi) = R_{n\ell}(r)\, Y_m^{\ell}(\theta, \phi)$) is

$$\frac{d^2}{dr^2} u_{n\ell}(r) = \left\{\frac{2m}{\hbar^2}(V(r) - E) + \frac{\ell(\ell+1)}{r^2}\right\} u_{n\ell}(r) , \tag{9.1}$$

with boundary conditions that $u_{n\ell}(0) = 0$, and $u_{n\ell}(r) \underset{r\to\infty}{\to} 0$.

Assuming $u_{n\ell}(r)$ is a real function one can readily show, integrating by parts that given a constant k

$$\int dr \frac{du_{n\ell}(r)}{dr} r^k u_{n\ell}(r) = -\frac{k}{2}\int dr\, u_{n\ell}(r) r^{k-1} u_{n\ell}(r) \equiv -\frac{k}{2} < r^{k-1} >, \tag{9.2}$$

(provided $r^k u_{n\ell}(r) \underset{\substack{r\to 0\\ r\to\infty}}{\to} 0$) while

$$\int dr \frac{du_{n\ell}(r)}{dr} r^k \frac{du_{n\ell}(r)}{dr} = \frac{k(k-1)}{2} < r^{k-2} > - \int dr\, u_{n\ell}(r)\, r^k \frac{d^2}{dr^2} u_{n\ell}(r)$$

$$= -\frac{2}{k+1}\int dr \frac{du_{n\ell}(r)}{dr} r^{k+1} \frac{d^2 u_{n\ell}(r)}{dr^2}$$

(provided

$$r^k \frac{du^2_{n\ell}(r)}{dr} \underset{\substack{r\to 0\\ r\to\infty}}{\to} 0 , \quad kr^{k-1} u^2_{n\ell}(r) \underset{\substack{r\to 0\\ r\to\infty}}{\to} 0, \quad \left(\frac{du_{n\ell}(r)}{dr}\right)^2 \frac{r^{k+1}}{k+1} \underset{\substack{r\to 0\\ r\to\infty}}{\to} 0, \; ,$$

and the integrals do not diverge), i.e.

$$\frac{k(k-1)}{2} < r^{k-2} > = \int dr \left\{ u_{n\ell}(r) r^k - \frac{2}{k+1} \frac{du_{n\ell}(r)}{dr} r^{k+1} \right\} \frac{d^2 u_{n\ell}(r)}{dr^2} . \tag{9.3}$$

Substituting Equation (9.1) into Equation (9.3) and using Equation (9.2) one obtains after regrouping terms:

$$\boxed{\frac{k\{k^2-1-4\ell(\ell+1)\}}{2(k+1)} < r^{k-2} > - \frac{2m}{\hbar^2}\left(\frac{2k+p+2}{k+1}\right) < Vr^k > + \frac{4mE}{\hbar^2} <r^k>=0} , \tag{9.4}$$

(provided additionally

$$\frac{\ell(\ell+1)}{k+1} r^{k-1} u^2_{n\ell}(r), \ \frac{r^{k+1}}{k+1} u^2_{n\ell}(r) \quad \text{and} \quad \frac{r^{k+p+1}}{k+1} u_{n\ell}(r)$$

go to zero as $r \to 0$ and $r \to \infty$). If one studies in detail the restrictions under which Equation (9.4) is valid it is obvious it is not valid if the constant $k = -1$. The constant k need not however be an integer.

If $k = 0$ Equation (9.4) reduces to:

$$E = \frac{p+2}{2} <V>. \tag{9.5}$$

But $E = <H> = <T> + <V>$. Thus for any quantum mechanical state $|n\ell>$ of H,

$$<n\ell|T|n\ell> = \frac{p}{2} <n\ell|V|n\ell>, \tag{9.6}$$

which is just the quantum mechanical analogue of the classical Virial Theorem[1]):

$$\bar{T} = \frac{p}{2} \bar{V}.$$

If $k = 1$,

$$-\ell(\ell+1) <r^{-1}> - \frac{2m(p+4)}{2\hbar^2} <rV> + \frac{4mE}{\hbar^2} <r> = 0,$$

i.e.

$$E = \frac{\frac{p+4}{4} <rV> + \frac{\hbar\ell(\ell+1)}{4m} < \frac{1}{r} >}{< r >} . \tag{9.7}$$

Equating expressions (9.5) and (9.7) one obtains

$$<rV> = \frac{2(p+2)}{p+4} <r> <V> - \frac{\hbar^2\ell(\ell+1)}{m(p+4)} < \frac{1}{r} >. \tag{9.8}$$

If

$$k = 2,$$

$$E = \frac{\frac{p+6}{6} <r^2V> + \frac{\hbar^2}{12m} (4\ell(\ell+1) - 3)}{<r^2>} \tag{9.9}$$

and

$$<r^2V> = \frac{3(p+2)}{p+6} <r^2> <V> - \frac{\hbar^2 (4\ell(\ell+1) - 3)}{2m(p+6)} \tag{9.10}$$

etc.

Equations (9.7) and (9.9) are natural generalizations of Equation (9.5), while Equation (9.8) and Equation (9.10) can be written as sum rules. Thus,

$$\sum_{n'\ell' \neq n\ell} <n\ell|V|n'\ell'><n'\ell'|r^2|n\ell> =$$

$$= \frac{2p}{p+6} <n\ell|V|n\ell><n\ell|r^2|n\ell> - \frac{\hbar^2(4\ell(\ell+1) - 3)}{2m(p+6)}, \tag{9.11}$$

with similar expressions for higher moments.

The above equations apply equally well to one-dimensional problems. For such problems $u_{n\ell}(r)$ becomes $\psi_N(x)$, the complete eigenfunction for the problem in question, while ℓ must be set equal to zero. That $\psi_N(x)$ has different boundary conditons, namely $\psi_N(x) \underset{x \to \pm\infty}{\to} 0$ merely implies the limits in the various integrals are x from $-\infty$ to ∞ in the one-dimensional case rather than r from 0 to ∞ and that the conditions are for instance

$$x^k \psi_N^2(x) \underset{x \to \pm\infty}{\to} 0 \qquad \text{etc.}$$

Thus:

$$\boxed{\frac{k(k-1)}{2} <x^{k-2}> - \frac{2m(2k+p+2)}{\hbar^2(k+1)} <x^kV> + \frac{4mE}{\hbar^2} <x^k> = 0.} \tag{9.12}$$

EXAMPLE 9.1. Consider the one-dimensional harmonic oscillator: $V(x) = \frac{1}{2} m\omega^2x^2$ (i.e. $p = 2$) where $E = (N + \frac{1}{2}) \hbar\omega$. Obtain expectation values of various powers of x. With these substitutions in Equation (9.12) one obtains:

$$\frac{k(k^2-1)}{2} <x^{k-2}> - \frac{m^2\omega^2(2k+4)}{\hbar^2(k+1)} <x^{k+2}> + \frac{4m\omega}{\hbar}(N+\frac{1}{2}) <x^k> = 0. \quad (9.13)$$

Substituting k = 0, 2, 4 in this equation yields:

$$<N|x^2|N> = \frac{\hbar}{m\omega}(N + \frac{1}{2})$$

$$<N|x^4|N> = \frac{3\hbar^2}{8m^2\omega^2}(4(N+\frac{1}{2})^2 + 1)$$

$$<N|x^6|N> = \frac{5\hbar^3}{12m^3\omega^3}(N+\frac{1}{2})\{6(N+\frac{1}{2})^2 + 7.5\} \quad (9.14)$$

etc.

EXAMPLES 9.2, 9.3. For two important three dimensional problems: $V(r) = -\alpha\hbar c/r$, $E = -\frac{1}{2}\alpha^2 mc^2 \frac{1}{n^2}$ (where $\alpha = e^2/(4\pi\varepsilon_o\hbar c) \simeq \frac{1}{137}$) and $V(r) = \frac{1}{2}m\omega^2 r^2$, $E_{n\ell} = (2n + \ell + 3/2)\hbar\omega$, obtain recursion relations.

Substituting into Equation (9.4) one gets for $V(r) = \alpha\hbar c/r$ Kramer's well known formula:

$$\frac{k(k^2-(2\ell+1)^2)}{4(k+1)} <r^{k-2}> + \frac{mc\alpha}{\hbar}\left(\frac{2k+1}{k+1}\right) < r^{k-1}> - \frac{m^2c^2\alpha^2}{n^2\hbar^2} <r^k> = 0, \quad (9.15)$$

and for $V(r) = \frac{1}{2}m\omega^2 r^2$

$$\frac{k(k^2-(2\ell+1)^2)}{4(k+1)} <r^{k-2}> + \frac{m^2\omega^2}{\hbar^2}\left(\frac{k+2}{k+1}\right) <r^{k+2}> + \frac{2m\omega}{\hbar}\left(2n+\ell+\frac{3}{2}\right)<r^k> = 0. \quad (9.16)$$

EXAMPLE 9.4. Show that if $V(x) = Ax^4$, $x > 0$; $V(x) = \infty$ $x < 0$

$$<n|xV|n> = 1.5<n|V|n> <n|x|n>.$$

In this case $0 < x < \infty$ in the integrals of Equation (9.12) though this is a one-dimensional problem.
Using Equation (9.12) with k = 1,

$$\frac{p+4}{2} < xV > = 2E <x>.$$

But from Equation (9.5) $E = (p+2)/2 \langle V \rangle$ (whether the system is one or three dimensional);
therefore

$$\langle xV \rangle = \frac{2(p+2)}{p+4} \langle V \rangle \langle x \rangle . \tag{9.17}$$

If $p = 4$

$$\langle xV \rangle = 1.5 \langle V \rangle \langle x \rangle .$$

EXAMPLE 9.5. For $V(x) = \alpha x^4$ show

$$\langle n|x^2 V|n\rangle = 1.8 \langle n|x^2|n\rangle \langle n|V|n\rangle + \frac{3\hbar^2}{20m} . \tag{9.18}$$

Substituting directly into Equation (9.10) with $p = 4$, $\ell = 0$

$$\langle r^2 V \rangle = 1.8 \langle r^2 \rangle \langle V \rangle + \frac{3\hbar^2}{20m} .$$

The same result applies to the one-dimensional case with $r \to x$.
If $V(x) = \alpha x^4$ for all x, the integrals extend over all x.
If $V(x) = \alpha x^4$ $x > 0$, $V(x) = \infty$ $x < 0$ the integrals extend over $x > 0$.

EXAMPLE 9.6. Given $V(x) = \alpha x$ $x > 0$, $V(x) = \infty$ $x < 0$,
show

$$(n|x^2|n) = \frac{6}{5} (n|x|n)^2 \tag{9.19}$$

$$(n|x^3|n) = \frac{9}{7}(n|x^2|n)\,(n|x|n) + \frac{3\hbar^2}{14m\alpha} . \tag{9.20}$$

Substituting in Equation (9.17) $p = 1$

$$\langle x\alpha x \rangle = \frac{6}{5} \langle \alpha x \rangle \langle x \rangle ,$$

which yields Equation (9.19), while substituting in Equation (9.10)
$p = 1$, $\ell = 0$ $r \to x$

$$\langle x^2 \alpha x \rangle = \frac{9}{7} \langle x^2 \rangle \langle \alpha x \rangle + \frac{3\hbar^2}{2m7} ,$$

which yields Equation (9.20).

EXAMPLE 9.7. If $V(x) = \alpha|x|$ show

$$\int_0^\infty \phi_n x^3 \phi_n \, dx = \frac{18}{7} \int_0^\infty \phi_n x^2 \phi_n \, dx \int_0^\infty \phi_n x \phi_n \, dx + \frac{3\hbar^2}{\alpha 28m} . \tag{9.21}$$

Substituting in Equation (9.10) $p = 1$, $\ell = 0$ and taking into account that $\phi_n^2(x) = \phi_n^2(-x)$ since $\phi_n(x)$ has a definite parity (because $V(x) = V(-x)$), one immediately gets Equation (9.21).

EXAMPLE 9.8. Obtain expressions for $x_{rms} = \sqrt{\langle N|x^2|N\rangle}$, if p = 1, 2, i.e. $V = A_1 x$ or $V = A_2 x^2$.

If p = 1, from Equation (9.17), $\langle N|x^2|N\rangle = 6/5A_1^2 \langle N|A_1 x|N\rangle^2$

$$\frac{6}{5A_1^2}\left(\frac{2}{3}E\right)^2 = \frac{8}{15}\frac{E^2}{A_1^2}$$

$$\therefore \quad x_{rms} = \sqrt{\frac{8}{15}}\frac{E}{A_1}\,. \tag{9.22}$$

If p = 2 from Equation (9.14),

$$\langle N|x^2|N\rangle = \hbar/m\omega \quad (N+\tfrac{1}{2}) = \frac{2}{m\omega^2}\frac{\hbar\omega}{2}(N+\tfrac{1}{2}) = \frac{E}{2\left(\frac{m\omega^2}{2}\right)} \quad \therefore$$

$$x_{rms} = \sqrt{\frac{1}{2}}\sqrt{\frac{E}{\frac{1}{2}m\omega^2}} = \sqrt{\frac{E}{2A_2}}\,. \tag{9.23}$$

EXAMPLE 9.9. Show if p = 2,

$$E_{n\ell} = \frac{(4\ell(\ell+1) - 3)\hbar^2}{4m}\frac{\langle n\ell|r^{-4}|n\ell\rangle}{\langle n\ell|r^{-2}|n\ell\rangle}\,. \tag{9.24}$$

Substituting k = -2 into Equation (9.4) immediately yields Equation (9.24). Note this result is independent of A, the constant of V(r). Since E > 0 $4\ell(\ell+1) > 3$ i.e. $\ell \geq 1$ for the integrals in Equation (9.24) to be convergent.

EXAMPLES 9.10, 9.11. Show

$$E_{n\ell} = (4\ell(\ell+1) - \frac{5}{4})\frac{3\hbar^2}{8m}\frac{\langle n\ell|r^{-7/2}|n\ell\rangle}{\langle n\ell|r^{-3/2}|n\ell\rangle} \tag{9.25}$$

if p = 1, and

$$E_{n\ell} = -(4\ell(\ell+1) + \frac{3}{4})\frac{\hbar^2}{8m}\frac{\langle n\ell|r^{-5/2}|n\ell\rangle}{\langle n\ell|r^{-1/2}|n\ell\rangle} \tag{9.26}$$

if p = -1.

Substituting k = -3/2 into Equation (9.26) yields result (9.25) and k = - 1/2 into Equation (9.4) yields result (9.26). Both these results are independent of A.

REFERENCE

H. Goldstein, Classical Mechanics, Addison-Wesley, (1950), p. 69.

CHAPTER 10

Upper Bounds and Parity Considerations

Consider a system with a Hamiltonian H such that

$$H\Psi_n = E_n\Psi_n. \tag{10.1}$$

If $\psi_t(\alpha, \beta)$ is a normalized 'trial' wavefunction with parameters $\alpha, \beta, \ldots,$ one can define the integral

$$E(\alpha, \beta) \equiv \int \psi_t^* H\psi_t \, dx. \tag{10.2}$$

Since the Ψ_n constitute a complete set one can expand $\psi_t(\alpha, \beta\ldots)$ in terms of the Ψ_n's in Equation (10.1) which are assumed normalized in what follows. Thus

$$\psi_t(\alpha, \beta) = \sum_{n=0}^{\infty} c_n\Psi_n, \tag{10.3}$$

where the fact that ψ_t and Ψ_n are normalized implies

$$\sum_{n=0}^{\infty} |c_n|^2 = 1. \tag{10.4}$$

Substituting Equation (10.3) into expression (10.2) yields

$$E(\alpha, \beta) = \int \sum_n c_n^* \Psi_n^* H \sum_m c_m \Psi_m \, dx =$$

$$= \sum_{n=0}^{\infty} |c_n|^2 E_n \geq E_0 \sum_{n=0}^{\infty} |c_n|^2 = E_0,$$

using Equation (10.4).

Hence

$$E(\alpha, \beta) \geq E_0 \,. \qquad (10.5)$$

This result enables one to use any trial wavefunction and in addition optimize, i.e. choose parameters α, β etc. which minimize $E(\alpha, \beta)$ for that particular trial wavefunction by requiring $\partial E(\alpha, \beta)/\partial\alpha = 0$ etc. In this way one gets upper bounds to the ground state of any quantum mechanical system.

The only restriction on $\psi_t(\alpha, \beta)$ is that it obeys the <u>same</u> boundary conditions as the eigenfunctions of H. Otherwise the assumption (10.3) is not valid. One also of course assumes the quantities one works with are well enough behaved that one can interchange summations and integrations in the expression for $E(\alpha, \beta)$.

EXAMPLE 10.1. Consider the system $H = T + V$ where

$$V = \frac{\hbar^2}{2mb^4} x^2 \qquad x > 0, \; V = \infty \qquad x < 0. \qquad (10.6)$$

The exact ground state wavefunction for this system and corresponding ground state energy are in fact known:

$$\Psi_0 = \left(\frac{1}{b\sqrt{\pi}}\right)^{\frac{1}{2}} \frac{2x}{b}\, e^{-x^2/2b^2} \; ; \qquad E_0 = \sqrt{2.25}\, \frac{\hbar^2}{mb^2}. \qquad (10.7)$$

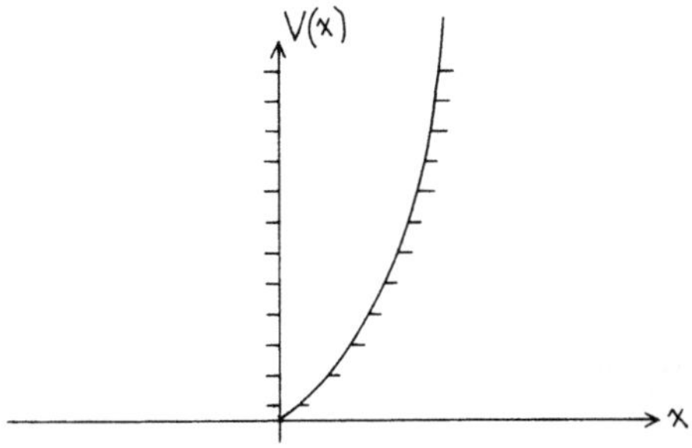

Figure 10.1. Potential in Example 10.1.

Consider the trial wavefunction

$$\psi_t = \sqrt{\frac{(2\alpha)^{2n+1}}{(2n)!}}\; x^n e^{-\alpha x} \,,$$

where n and α are parameters, $n > 0$.
Performing the integral (10.2) one obtains

$$E(\alpha, n) = \frac{\hbar^2\alpha^2}{(2n-1)2m} + \frac{(2n+1)(n+1)\hbar^2}{(2\alpha)^2 mb^4} .$$

$$\frac{\partial E}{\partial \alpha} = 0 \text{ implies } \quad \alpha^4 = \frac{(4n^2-1)(n+1)}{2b^4}$$

while $\partial E/\partial n = 0$ implies $4\alpha^4 b^4 = (4n+3)(2n-1)^2$

i.e.

$$E(n) = \sqrt{\frac{(2n+1)(n+1)}{2(2n-1)}} \; \frac{\hbar^2}{mb^2} , \tag{10.8}$$

with optimal $n = (1 + \sqrt{6})/2$.
If $n = 1$ $E(1) = \sqrt{3}\ \hbar^2/mb^2$. If $n = 1.5$ or 2, $E(1.5) = E(2) = \sqrt{2.5}\ \hbar^2/mb^2$,
and if one chooses the optimal $n = (1 + \sqrt{6})/2$ one obtains
$E((1 + \sqrt{6})/2) = \sqrt{2.47}\ \hbar^2/mb^2$, which is quite close but slightly larger than the exact result Equation (!0.7) (as it must be).

EXAMPLE 10.2. Consider the system $H = T + V$ where

$$V = Ax \quad x > 0, \qquad V = \infty \quad x < 0, \qquad A > 0.$$

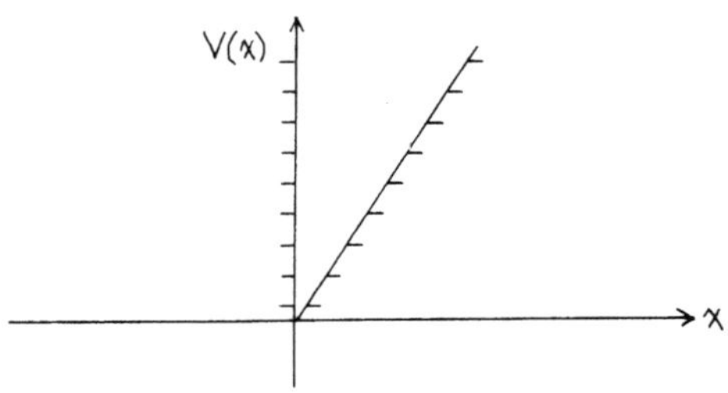

Figure 10.2. Potential in Example 10.2.

If one uses the trial wavefunction

$$\psi_t^{(1)} = \left(\frac{1}{b\sqrt{\pi}}\right)^{\frac{1}{2}} \frac{2x}{b}\ e^{-x^2/2b^2},$$

where b is a free parameter,

$$E(b) = \frac{3\hbar^2}{4mb^2} + \frac{2Ab}{\sqrt{\pi}} .$$

The requirement $\partial E/\partial b = 0$ implies

$$b^3 = \frac{3\sqrt{\pi}\hbar^2}{4mA} .$$

This yields

$$E(b_{optimal}) = \left(\frac{81}{4\pi}\right)^{1/3} \left[\frac{A^2\hbar^2}{m}\right]^{1/3} \sim 1.86 \left[\frac{A^2\hbar^2}{m}\right]^{1/3} .$$

Consider the trial wavefunction

$$\psi_t^{(2)} = \sqrt{\frac{(2\alpha)^{2n+1}}{(2n)!}}\; x^n e^{-\alpha x}$$

where n and α are arbitrary parameters.

This function, when substituted in expression (10.2) yields

$$E(\alpha, n) = \frac{\hbar^2\alpha^2}{(2n-1)2m} + \frac{(2n+1)A}{2\alpha}$$

$$\frac{\partial E}{\partial \alpha} = 0 \text{ implies } \alpha^3 = \frac{(4n^2-1)Am}{2\hbar^2}$$

$$\frac{\partial E}{\partial n} = 0 \text{ implies } \alpha^3 = \frac{Am(2n-1)^2}{\hbar^2}$$

Substituting in $E(\alpha, n)$ yields

$$E(n) = \left(\frac{27(2n+1)^2}{32(2n-1)}\right)^{1/3} \left(\frac{\hbar^2A^2}{m}\right)^{1/3} \text{ with optimal } n = 1.5. \qquad (10.9)$$

If

$$n = 1, \; E(1) = (1.5)^{5/3}\left(\frac{\hbar^2A^2}{m}\right)^{1/3} \approx 1.97\left(\frac{\hbar^2A^2}{m}\right)^{1/3} .$$

If

$$n = 2, \; E(2) = \left(\frac{225}{32}\right)^{1/3} \left(\frac{\hbar^2A^2}{m}\right)^{1/3} \approx 1.92\left(\frac{\hbar^2A^2}{m}\right)^{1/3} .$$

Finally if one chooses the optimal n = 1.5

$$E(1.5) = \left(\frac{27}{4}\right)^{1/3}\left(\frac{\hbar^2A^2}{m}\right)^{1/3} \approx 1.89\left(\frac{\hbar^2A^2}{m}\right)^{1/3} .$$

It is interesting to note that even with the best two parameter trial wavefunction $\psi_t^{(2)}$ one does not do as well as with the one parameter trial wavefunction $\psi_t^{(1)}$. In other words increasing the parameters in one's wavefunction <u>does not necessarily</u> result in lower energies i.e. better results.

Consider the trial wavefunction $\psi_t^{(3)} = \left(\frac{2}{b\sqrt{\pi}}\right)^{1/2} e^{-x^2/2b^2}$ with b as a free parameter.

$$E(b) = \frac{\hbar^2}{4mb^2} + \frac{Ab}{\pi^{\frac{1}{2}}} . \quad \frac{\partial E}{\partial b} = 0 \text{ implies } b^3 = \frac{\hbar^2\pi^{\frac{1}{2}}}{2mA} .$$

Substituting this value of b into E(b) implies

$$E(b_{optimal}) = \left(\frac{27}{16\pi}\right)^{1/3}\left(\frac{A^2\hbar^2}{m}\right)^{1/3} \approx 0.81\left(\frac{A^2\hbar^2}{m}\right)^{1/3} .$$

This is much less than $1.86\left(A^2\hbar^2/m\right)^{1/3}$, the best value obtained with the other two trial wavefunctions! But this wavefunction <u>does not</u> satisfy the boundary conditions and hence is unacceptable. In particular it is not zero at x = 0. Hence this particular result is wrong!

The lowest energy for large arguments of the relevant Airy function is $b = 1.84\left(A^2\hbar^2/m\right)^{1/3}$ (see Equation (3.18)).

EXAMPLE 10.3. Consider the system H = T + V where $V = A|x|$ (A > 0) for all x.

If P, the so called 'parity' operator is such that when operating on any function f(x)

$$P\, f(x) = f(-x),$$

Then $PH(x)\psi(x) = H(-x)\psi(-x) = H(-x)P\psi(x) = H(x)P\psi(x)$ for this case since

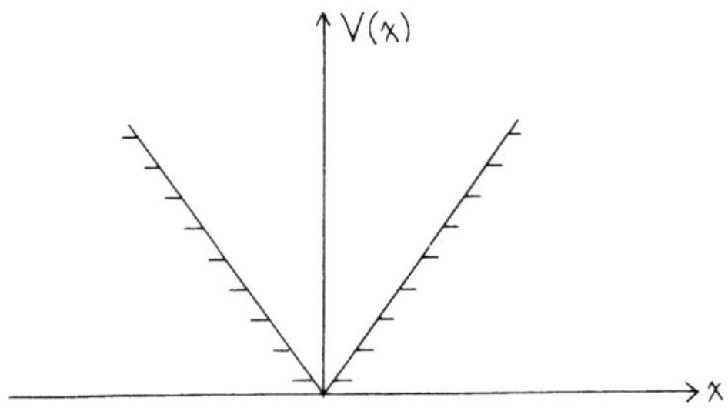

Figure 10.3. Potential in Example 10.3.

for the V of this example $V(x) = V(-x)$ while quite generally $T(x) = T(-x)$. Hence in this case $H(x) = H(-x)$. Thus for this problem (indeed whenever $V(x) = V(-x)$)

$$[PH(x) - H(x)P]\psi(x) = 0 \quad \text{i.e.} \quad [P, H] = 0.$$

But this has consequences on the eigenvalues and eigenfunctions of H. Assuming $[H, P] = 0$ consider the Schrödinger equation (10.1)

$H\Psi_n(x) = E_n \Psi_n(x)$. Premultiplying by P

$$PH(x)\Psi_n(x) = H(x)P\Psi_n(x) = E_n P\Psi_n(x).$$

Assuming the system is non-degenerate (see Chapter 12), if Ψ_n is an eigenfunction of H and at the same time $P\Psi_n$ is an eigenfunction of H with the same eigenvalue E_n, this implies

$$P\Psi_n(x) = \lambda\Psi_n(x). \tag{10.10}$$

i.e. $\Psi_n(x)$ is proportional to $\Psi_n(x)$.

But Equation (10.10) is just the eigenvalue equation for P. The eigenfunctions of the Hamiltonian are also eigenfunctions of P! Consider now the eigenvalue problem for P. If p_0 are the eigenvalues of P,

$$\begin{aligned} Pf(x) &= p_0 f(x) \\ &= f(-x) \\ P^2 f(x) &= p_0^2 f(x) = f(x). \end{aligned} \tag{10.11}$$

Hence $p_0 = \pm 1$.

Thus in Equation (10.10) $\lambda = \pm 1$ and $\Psi_n(-x) = \Psi_n(x)$ or

$$\Psi_n(-x) = -\Psi_n(x).$$

Thus whenever, as is the case in this problem $V(x) = V(-x)$ and the system is nondegenerate, the eigenfunctions of H have either even parity or odd parity.

Going through the derivation for upper bounds (1) - (5) one sees that in this case Equation (10.5) applies <u>independently</u> to the odd and even parity solutions since Equation (10.3) will be an expansion either in terms of the even or the odd eigenfunctions Ψ_n. Hence working with odd parity trial wavefunctions one gets an upper bound to the lowest odd parity state energy and similarly for even parity trial wavefunctions an upper bound to the lowest even parity state energy.

A simple (though inadequate) even parity trial wavefunction for Example 10.3 is:

$$\psi_t(b) = \sqrt{\frac{3}{2b^3}}\,(b-|x|) \qquad |x| < b$$

$$= 0 \qquad |x| > b$$

with free parameter b, yielding,

$$E^+(b) = \frac{3\hbar^2}{2mb^2} + \frac{Ab}{4} \;;\quad \frac{\partial E^+}{\partial b} = 0 \text{ implies } b^3 = \frac{12\hbar^2}{Am}$$

(where one uses the representation of the delta function

$$\frac{d^2}{dx^2}\,|x| = 2\delta(x)).$$

For optimal

$$b,\; E_g^+ \le E_g^+(b_{opt}) = \frac{4.5}{12^{2/3}}\left(\frac{A^2\hbar^2}{m}\right)^{1/3} \doteq 0.86\left(\frac{A^2\hbar^2}{m}\right).$$

A better <u>even</u> parity trial wavefunction one can use is

$$\psi_t^+(b) = \sqrt{\frac{1}{b\sqrt{\pi}}}\; e^{-x^2/2b^2}$$

with free parameter b, yielding

$$E^+(b) = \frac{\hbar^2}{4mb^2} + A\,\frac{b}{\sqrt{\pi}} \;;\text{ if } \frac{\partial E^+}{\partial b} = 0,\; b^3 = \frac{\hbar^2\pi^{\frac{1}{2}}}{2mA}\;.$$

For optimal

$$b,\, E_g^+ \le E_g^+(b_{opt}) = \left(\frac{27}{16\pi}\right)^{1/3}\left(\frac{A^2\hbar^2}{m}\right)^{1/3} \doteq 0.81\left(\frac{A^2\hbar^2}{m}\right)^{1/3}.$$

An <u>odd</u> parity trial wavefunction one can use is

$$\psi_t = \left(\frac{2}{b\sqrt{\pi}}\right)^{1/2} \frac{x}{b}\; e^{-x^2/2b^2}$$

with free parameter b, yielding

$$E^-(b) = \frac{3\hbar^2}{4mb^2} + \frac{2Ab}{\sqrt{\pi}}. \quad \text{If} \quad \frac{\partial E^-}{\partial b} = 0,\; b^3 = \frac{3\hbar^2\pi^{1/2}}{4mA},$$

and for optimal b:

$$E_g^- \le E^-(b_{optimal}) = 1.86\left(\frac{\hbar^2 A^2}{m}\right)^{1/3} .$$

One thus has roughly determined two energy levels with the help of parity considerations in this case. One notes

$$\frac{E_g^-(\text{opt})}{E_g^+(\text{opt})} = \frac{1.86}{0.81} \approx 2.29$$

as opposed to

$$\frac{E_g^-(\text{exact})}{E_g^+(\text{exact})} = \frac{1.84}{0.89} \approx 2.07$$

(cf. Equation (3.18), (3.31)).

EXAMPLE 10.4. Consider the system:

$$H = \frac{p^2}{2m} - V_0\delta(x) \quad V_0 > 0.$$

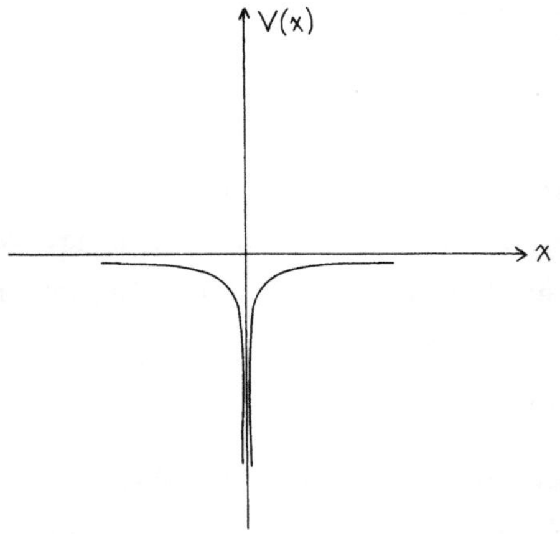

Figure 10.4. Potential in Example 10.4.

Estimate the energy of this system using as trial wavefunction

$$\psi_t = \left(\frac{m\omega}{\hbar\pi}\right)^{1/4} e^{-\frac{m\omega x^2}{2\hbar}}$$

(the groundstate wavefunction of the one-dimensional harmonic oscillator), with free parameter ω. One obtains directly

$$E(\omega) = \frac{\hbar\omega}{4} - V_0\left(\frac{m\omega}{\hbar\pi}\right)^{1/2} .$$

Requiring

$$\frac{\partial E(\omega)}{\partial\omega} = 0 \quad \text{implies} \quad \frac{\hbar}{4} - \frac{V_0}{2}\left(\frac{m}{\hbar\pi}\right)^{1/2} \frac{1}{\omega^{1/2}} = 0$$

i.e.

$$\omega = \frac{4mV_0^2}{\pi\hbar^3} .$$

Hence for this trial wavefunction,

$$E_{\text{optimal}} = \frac{mV_0^2}{\pi\hbar^2} - V_0\left(\frac{m}{\hbar\pi}\right)^{1/2} 2\left(\frac{mV_0^2}{\pi\hbar^3}\right)^{1/2} = -\frac{mV_0^2}{\hbar^2\pi} ,$$

vs the exact ground (and only) state and only energy of this system (cf. Equation (13.11)) $= -mV_0^2/2\hbar^2$.

One again notes

$$E_{\text{optimal}} = -0.318\frac{mV_0^2}{\hbar^2} > Eg = -0.5\frac{mV_0^2}{\hbar^2}$$

as required by Equation (10.5).

EXAMPLE 10.5. Consider the system $H = p^2/2m + V$ where $V = \frac{1}{2}m\omega^2x^2 + \alpha/x^2 \quad x \geq 0, \quad V = \infty \quad x < 0,$ (α positive or negative).

Estimate the ground-state energy of this system using as trial wavefunction:

$$\psi_t^{(1)} = \left(\frac{m\omega'}{\hbar}\right)^{3/4} \frac{2x}{\pi^{1/4}} e^{-\frac{m\omega' x^2}{2\hbar}}$$

(This is the lowest eigenfunction of $H' = p^2/2m + V'$ where $V' = \frac{1}{2}m\omega'^2x^2 \quad x \geq 0, \quad V' = \infty \quad x < 0,$ with eigenvalue $\frac{3}{2}\hbar\omega'$.)

Using this trial wavefunction,

$$E_t^{(1)}(\omega') = \frac{3}{2}\hbar\omega' + \frac{2m\omega'\alpha}{\hbar} + \left(\frac{\omega^2}{\omega'^2} - 1\right)\frac{3}{4}\hbar\omega' .$$

(This result can easily be obtained by rewriting

$$H = \frac{p^2}{2m} + \frac{1}{2}m\omega'^2 x^2 + \frac{\alpha}{x^2} + \left(\frac{\omega^2}{\omega'^2} - 1\right)\frac{1}{2}m\omega'^2x^2$$

and noting that

$$\int_0^\infty \psi_t^{(1)*} \frac{\alpha}{x^2} \psi_t^{(1)} dx = \frac{2m\omega'\alpha}{\hbar} ; \int_0^\infty \psi_t^{(1)*} \frac{1}{2}m\omega'^2 x^2 \psi_t^{(1)} dx =$$

$$= \frac{1}{2}\left(\frac{3}{2}\hbar\omega'\right) \text{(cf. Equation (9.5))}$$

From the constraint $(\partial E_t^{(1)}(\omega'))/\partial\omega' = 0$ one obtains

$$\omega'_{optimal} = \frac{\omega}{\sqrt{1 + \frac{8m\alpha}{3\hbar^2}}} .$$

Hence

$$E_t^{(1)}(\omega'=\omega_{optimal}) = \frac{3}{2}\hbar\omega \frac{\left\{1+\frac{4}{3}\frac{m\alpha}{\hbar^2}\right\}}{\sqrt{1 + \frac{8m\alpha}{3\hbar^2}}} + \frac{2m\omega\alpha}{\hbar\sqrt{1 + \frac{8m\alpha}{3\hbar^2}}} .$$

If $\alpha = \hbar^2/m$ the Hamiltonian above admits of an exact solution since $\hbar^2/mr^2 = 2\hbar^2/2mr^2$ is then just the centripetal term in the three dimensional simple harmonic oscillator ($V = \frac{1}{2}m\omega^2r^2$) Hamiltonian, if $\ell = 1$. The exact energies in this case are $(2n + 1 + \frac{3}{2})\hbar\omega$ $n = 0, 1, 2...$ i.e. the lowest solution is $E_0 = \frac{5}{2}\hbar\omega$.

In this case

$$E_t^{(1)}(\omega_{optimal}) = \frac{\sqrt{33}}{2}\hbar\omega \sim 2.873\ \hbar\omega > 2.5\ \hbar\omega$$

as expected.

One may use instead:

$$\psi_t^{(2)} = \sqrt{\frac{(2\beta)^{2n+1}}{(2n)!}}\, x^n e^{-\beta x}$$

as trial wavefunction.

In this case

$$E_t^{(2)}(\beta, n) = \frac{\hbar^2}{2m(2n-1)} \left\{\frac{n+4}{n}\right\} \beta^2 + \frac{m\omega^2(2n+2)(2n+1)}{8\beta^2} .$$

For optimal

$$\beta = \left(\frac{m\omega}{2\hbar}\right)^{1/2} \left(\frac{(2n-1)n(2n+1)(2n+2)}{n+4}\right)^{1/4}$$

$$E_t^{(2)}(\beta_{optimal}, n) = \hbar\omega \sqrt{\frac{(2n+1)(n+1)(n+4)}{2n(2n-1)}}$$

which has values 2.556 $\hbar\omega$ and 2.535 $\hbar\omega$ for n = 3 and 4 respectively, while for optimal $n \sim 3.74$, $E(\beta_{opt}, n_{opt}) \sim 2.533\ \hbar\omega > 2.5\ \hbar\omega$. Again as expected one gets an energy greater than the exact energy but $\psi_t^{(2)}$ gives a better upper bound than $\psi_t^{(1)}$.

If $\alpha = 3\hbar^2/m$ again one knows the exact solution which one can compare with the upper bound results that arise in this case etc. Thus for $\alpha = 3\hbar^2/m$, $E_0 = \frac{7}{2}\ \hbar\omega$, while $E_t^{(1)}(\omega_{optimal}) = 4.5\ \hbar\omega > 3.5\ \hbar\omega$. If

$$\psi_t^{(3)} = \left(\frac{8}{3\sqrt{\pi}}\right)^{1/2} \left(\frac{m\omega'}{\hbar}\right)^{5/4} x^2 e^{-m\omega' x^2/2\hbar}$$

is used as trial wavefunction,

$$E_t^{(3)}(\omega') = \left(\frac{7}{12} + \frac{2\alpha m}{3\hbar^2}\right)\hbar\omega' + \frac{5}{4}\hbar\frac{\omega^2}{\omega'} .$$

With

$$\omega'_{optimal} = \frac{\omega}{\sqrt{\frac{7}{15} + \frac{8\alpha m}{15\hbar^2}}} .$$

$$E_t^{(3)}(\omega'_{opt}) = \left(\frac{7}{12} + \frac{2\alpha m}{3\hbar^2}\right) \frac{\hbar\omega}{\sqrt{\frac{7}{15} + \frac{8\ \alpha m}{15\hbar^2}}} + \frac{5}{4}\,\hbar\omega\sqrt{\frac{7}{15} + \frac{8}{15}\frac{\alpha m}{\hbar^2}} .$$

If

$$\alpha = \frac{\hbar}{m} ,\ E_t^{(3)}(\omega'_{opt}) = 5/2\hbar\omega = E_{exact} .$$

If

$$\alpha = \frac{3\hbar^2}{m} , E_t^{(3)}(\omega'_{opt}) = \frac{1}{2}\sqrt{\frac{155}{3}}\ \hbar\omega = 3.59\ \hbar\omega > \frac{7}{2}\ \hbar\omega = E_{exact} .$$

EXAMPLE 10.6. Consider the system $H = p^2/2m + V$ where $V = -\alpha\hbar c/x$ $x \geq 0$, $= \infty$ $x < 0$, with groundstate eigenvalue of $-0.5mc^2\alpha^2$.

Estimate this ground state energy using as trial wavefunction:

$$\psi_t^{(1)} = \frac{2}{b}\left(\frac{1}{b\sqrt{\pi}}\right)^{1/2} xe^{-x^2/2b^2}.$$

Performing the integral in Equation (10.2) one obtains:

$$E^{(1)}_{(b)} = \frac{3\hbar^2}{4mb^2} - \frac{2\hbar c\alpha}{b\sqrt{\pi}} .$$

$$\frac{\partial E^{(1)}}{\partial b} = 0 \quad \text{implies } b_{opt} = \frac{3\hbar\sqrt{\pi}}{4mc\alpha} .$$

Hence

$$E^{(1)}(b_{optimal}) = -\frac{4}{3\pi} mc^2\alpha^2 \sim (-0.424\ mc^2\alpha^2)$$

$$> E_{ground} \quad \text{as it must be.}$$

This result is quite close to the exact ground-state energy. Other trial wavefunctions do not do as well.

Thus if instead one uses:

$$\psi_t^{(2)} = \frac{2}{b^2}\sqrt{\frac{2}{3b\sqrt{\pi}}}\ x^2e^{-x^2/2b^2} ,$$

one obtains

$$E^{(2)}(b) = 7\hbar^2/(12mb^2) - 4\hbar c\alpha/(3b\sqrt{\pi}).$$

$$\frac{\partial}{\partial b} E^{(2)}(b) = 0 \text{ implies } b_{opt} = \frac{7\hbar\sqrt{\pi}}{8mc\alpha} .$$

Hence

$$E^{(2)}(b_{optimal}) = -\frac{16}{21\pi} mc^2\alpha^2 \sim -0.243\ mc^2\alpha^2.$$

If one uses:

$$\psi_t^{(3)} = \sqrt{\frac{(2b)^5}{4!}}\ x^2e^{-bx}$$

$$E^{(3)}(b) = \frac{\hbar^2b^2}{6m} - \frac{\alpha\hbar cb}{2}.$$

$$\frac{\partial E^{(3)}}{\partial b} = 0 \quad \text{implies } b_{opt} = \frac{3\alpha mc}{2\hbar} .$$

Hence

$$E^{(3)}(b_{optimal}) = -\frac{3}{8} mc^2\alpha^2 = -0.375\ mc^2\alpha^2.$$

Though both $E^{(2)}(b_{opt})$ and $E^{(3)}(b_{opt})$ are above E_{ground}, neither is as close to it as $E^{(1)}(b_{opt})$.

A fourth trial wavefunction:

$$\psi_t^{(4)} = \frac{\sqrt{2}}{b^2} x^{3/2} e^{-x^2/2b^2}$$

yields:

$$E^{(4)}(b) = \frac{5\hbar^2}{8mb^2} - \frac{c\hbar c\sqrt{\pi}}{2b}, \quad b_{opt} = \frac{5\hbar}{2m\alpha c\sqrt{\pi}}$$

and

$$E^{(4)}(b_{opt}) = -\frac{\pi}{10} mc^2\alpha^2 \sim -0.31\ mc^2\alpha^2.$$

CHAPTER 11

Perturbation Theory

Consider a system whose Hamiltonian H contains a 'perturbation' V such that

$$H = H_0 + V. \tag{11.1}$$

If one knows the eigenfunctions and eigenvalues of H_0, i.e.

$$H_0\phi_n = \varepsilon_n\phi_n, \tag{11.2}$$

and wishes to find the eigenvalues and eigenfunctions of H i.e. E_n, ψ_n in

$$H\psi_n = E_n\psi_n, \tag{11.3}$$

One can formally expand each eigenfunction ψ in terms of the complete set ϕ_n:

$$\psi = \sum_{m=0}^{\infty} a_m\phi_m \tag{11.4}$$

where the ϕ's are assumed orthonormal i.e. $\int \phi_p^*\phi_q \, d\tau = \delta_{pq}$.

One can then rewrite Equation (11.3) as

$$\sum_{m=0}^{\infty} (H-E)\, a_m\phi_m = 0 , \tag{11.5}$$

where for notational simplicity the subscript n has been omitted from the ψ's and E's.

Premultiplying expression (11.5) by ϕ_p^* and integrating over the relevant variables one obtains:

$$\sum_{m=0}^{\infty} (H_{pm} - E\delta_{pm})a_m = 0, \qquad p = 0, 1, \ldots \tag{11.6}$$

where

$$H_{pm} = \varepsilon_m \delta_{pm} + V_{pm},$$

and

$$V_{pm} \equiv \int \phi_p^* V \phi_m \, d\tau = (\phi_p|V|\phi_m).$$

written in matrix form one has:

$$\begin{bmatrix} H_{00}-E & V_{01} & V_{02} \cdots \\ V_{10} & H_{11}-E & V_{12} \cdots \\ V_{20} & V_{21} & H_{22}-E \cdots \end{bmatrix} \begin{bmatrix} a_0 \\ a_1 \\ a_2 \end{bmatrix} = 0. \tag{11.7}$$

The eigenvalues E of Equation (11.7) are the exact energies of the system (11.3) and the corresponding coefficients a_i give, when substituted into Equation (11.4) the corresponding eigenfunctions.

Consider for simplicity the special case when expression (11.7) is a 2 × 2 matrix. The determinant of this matrix must be zero i.e.

$$E^2 - E(H_{00} + H_{11}) - V_{01}V_{10} = 0, \tag{11.8}$$

hence

$$E = \frac{(H_{00}+ H_{11}) \pm (H_{00}-H_{11})\left(1 + \dfrac{4V_{01}V_{10}}{(H_{00}-H_{11})^2}\right)^{1/2}}{2} \tag{11.9a}$$

$$\simeq \frac{(H_{00}+H_{11}) \pm (H_{00}-H_{11})\left(1 + \dfrac{2V_{01}V_{10}}{(H_{00}-H_{11})^2}\right)}{2} . \tag{11.9b}$$

This gives two energies:

$$E_0 \doteq H_{00} + \frac{V_{01}V_{10}}{H_{00}-H_{11}} = \varepsilon_0 + V_{00} + \frac{V_{01}V_{10}}{\varepsilon_0+V_{00}-\varepsilon_1-V_{11}} = \varepsilon_0+V_{00} + \frac{V_{01}V_{10}}{(\varepsilon_0-\varepsilon_1)\left(1+\dfrac{V_{00}-V_{11}}{\varepsilon_0-\varepsilon_1}\right)}$$

$$E_1 \doteq H_{11} + \frac{V_{10}V_{01}}{H_{11}-H_{00}} = \varepsilon_1 + V_{11} + \frac{V_{10}V_{01}}{\varepsilon_1+V_{11}-\varepsilon_0-V_{00}} = \varepsilon_1+V_{11} + \frac{V_{01}V_{10}}{(\varepsilon_1-\varepsilon_0)\left(1+\dfrac{V_{11}-V_{00}}{\varepsilon_1-\varepsilon_0}\right)}$$

(11.10)

Upon expanding the denominators in these expressions one obtains:

$$E_0 = \varepsilon_0+V_{00} + \frac{V_{01}V_{10}}{\varepsilon_0-\varepsilon_1} + \frac{V_{01}V_{11}V_{10}}{(\varepsilon_0-\varepsilon_1)^2} - V_{00}\frac{V_{01}V_{10}}{(\varepsilon_0-\varepsilon_1)^2} + \ldots$$

$$E_1 = \varepsilon_1+V_{11} + \frac{V_{10}V_{01}}{\varepsilon_1-\varepsilon_0} + \frac{V_{10}V_{00}V_{01}}{(\varepsilon_1-\varepsilon_0)^2} - V_{11}\frac{V_{10}V_{01}}{(\varepsilon_1-\varepsilon_0)^2} + \ldots \quad (11.11)$$

From the 2 × 2 matrix one can also obtain expressions for ψ by solving for the eigenvectors a_0 and a_1,

$$\psi_0 = a_0\left\{\phi_0 + \frac{E-H_{00}}{V_{01}}\phi_1\right\} \rightarrow \left\{\phi_0 + \frac{E-H_{00}}{V_{01}}\phi_1\right\}, \quad (11.12a)$$

and

$$\psi_1 = a_1\left\{\frac{E-H_{11}}{V_{10}}\phi_0 + \phi_1\right\} \rightarrow \left\{\frac{E-H_{11}}{V_{10}}\phi_0 + \phi_1\right\}, \quad (11.12b)$$

if in addition one requires $(\phi_0|\psi_0) = (\phi_1|\psi_1) = 1$ i.e. that the ψ's are 'cross normalized' functions.

Substituting E_0 and E_1 of Equation (11.11) into Equation (11.12a) and Equation (11.12b) respectively yields:

$$\psi_0 = \phi_0 + \frac{\phi_1 V_{10}}{\varepsilon_0-\varepsilon_1} + \frac{\phi_1 V_{11}V_{10}}{(\varepsilon_0-\varepsilon_1)^2} - \frac{\phi_1 V_{10}V_{00}}{(\varepsilon_0-\varepsilon_1)^2} + \ldots \quad (11.13)$$

$$\psi_1 = \phi_1 + \frac{\phi_0 V_{01}}{\varepsilon_1-\varepsilon_0} + \frac{\phi_0 V_{00}V_{01}}{(\varepsilon_1-\varepsilon_0)^2} - \frac{\phi_0 V_{01}V_{11}}{(\varepsilon_1-\varepsilon_0)^2} + \ldots \,.$$

One can generalize the results in Equation (11.11) as follows:

$$E_n = \varepsilon_n+V_{nn} + \sum_{m\neq n}\frac{V_{nm}V_{mn}}{\varepsilon_n-\varepsilon_m} + \left\{\sum_{m,p\neq n}\frac{V_{nm}V_{mp}V_{pn}}{(\varepsilon_n-\varepsilon_m)(\varepsilon_n-\varepsilon_p)} - V_{nn}\sum_{m\neq n}\frac{V_{nm}V_{mn}}{(\varepsilon_n-\varepsilon_m)^2}\right\}, \ldots \quad (11.14)$$

which is the standard Rayleigh-Schrödinger expansion for the exact energy to third order in V. The first term in Equation (11.14) is called zeroth-order. Terms in Equation (11.14) involving V once (the second term) are called first order. Terms involving V twice (the third term) are called second order, and terms involving V thrice (the fourth and fifth terms) are called third order.

One can generalize the results in Equation (11.13) as follows:

$$\psi_n = \phi_n + \sum_{m\neq n} \frac{\phi_m V_{mn}}{\varepsilon_n - \varepsilon_m} + \sum_{m,\, p\neq n} \frac{\phi_m V_{mp} V_{pn}}{(\varepsilon_n-\varepsilon_m)(\varepsilon_n-\varepsilon_p)} -$$

$$- V_{nn} \sum_{m\neq n} \frac{\phi_m V_{mn}}{(\varepsilon_n-\varepsilon_m)^2} + \ldots \qquad (11.15)$$

which is the standard Rayleigh-Schrödinger expression for the exact wave-function to second-order in perturbation theory when one has cross-normalized functions ψ_n i.e. $(\phi_n|\psi_n) = 0$. Fourth and higher-order terms and third and higher-order terms in Equation (11.11) and (11.13) respectively may be obtained by keeping more terms in going from Equation (11.9a), to Equation (11.9b). The usefulness of expression (11.14) and (11.15) in turn depends among other things on whether the expansions converge.

EXAMPLE 11.1. Consider a particle subject to the Hamiltonian

$$H = T + \frac{1}{2} m\omega^2 x^2 \quad |x| < a, \qquad H = T + \frac{1}{2} m\omega^2 a^2 \quad |x| > a.$$

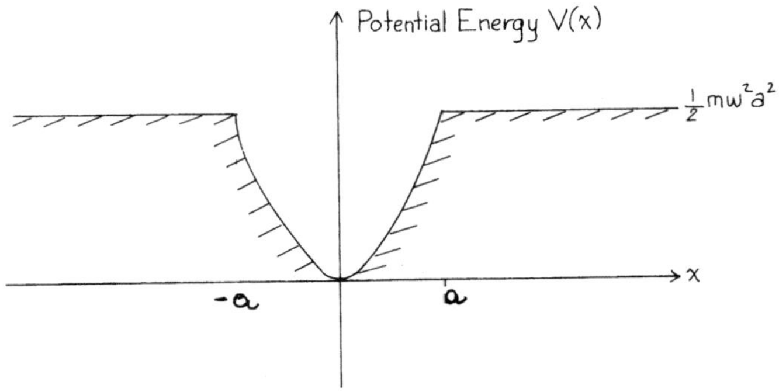

Figure 11.1. Potential Energy in Example 11.1.

One can rewrite this Hamiltonian as

$$H = H_0 + V \quad \text{where} \quad H_0 = T + \frac{1}{2} m\omega^2 x^2$$

$$V = 0 \qquad |x| < a$$

$$= -\frac{1}{2} m\omega^2 (x^2 - a^2) \qquad |x| > a.$$

To first order in V one then has

$$E_n = \varepsilon_n + V_{nn}$$
$$= \varepsilon_n - \frac{1}{2} m\omega^2 \, 2\int_a^\infty (x^2-a^2)\phi_n^2(x)\, dx$$

where ε_n, $|n\rangle$ are the standard eigenvalues and eigenfunctions of the one-dimensional infinite harmonic oscillator.

This problem illustrates one important limitation of perturbation theory. It predicts here an infinite, discrete set of energies E_n. However, if $E > \frac{1}{2} m\omega^2 a^2$ the particle is no longer bound but rather is free i.e. it can have any energy!

In fact only if $E \ll \frac{1}{2} m\omega^2 a^2$ does one expect to get reasonably accurate results using perturbation theory.

Classically for this system it makes no difference what the potential is for x greater than x_{max} where $x_{max} = \sqrt{2E/m\omega^2}$.

The quantum mechanical treatment of this problem however, shows that the potential for $x < x_{max}$ affects the particle. If $a \to \infty$ the exact energies of the system go to E_n. For the actual potential however, the exact energies E_n are less than ε_n since $V_{nn} < 0$. Roughly speaking this is because there is more likelyhood the particle will be in the classically forbidden region if $V = \frac{1}{2} m\omega^2 a^2$ for $x > a$ than if it is more repulsive i.e. $\frac{1}{2} m\omega^2 x^2$ in this region, and the more the particle spreads the bigger its wavelength λ and smaller its energy since $k \sim 1/\lambda$ while $E \sim k^2$.

EXAMPLE 11.2. Consider a particle in the potential

$$V(x) = Ax \quad x > 0 \quad (A > 0), \quad V(x) = \infty \quad x < 0.$$

Suppose one wishes to know the ground state energy of a particle in this potential.

The Hamiltonian of the system is $H = T + Ax$, $x > 0$. But one can rewrite H as

$$H = H_0 + (Ax - \frac{\hbar^2}{2mb^4} x^2)$$

where

$$H_0 = T + \frac{\hbar^2}{2mb^4} x^2 \quad (x > 0),$$

and b is a parameter.

One can then treat $Ax - \hbar^2/2mb^4\, x^2$ $(x > 0)$ as the perturbation V. For this choice of H_0,

$$\phi_0 = \frac{2x}{b}\left(\frac{1}{b\sqrt{\pi}}\right)^{1/2} e^{-x^2/2b^2} = |0),$$

and

$$\varepsilon_0 = (\phi_0|H_0|\phi_0) = \frac{3\hbar^2}{2mb^2} .$$

The first order energy is

$$\int \phi_0^*\left(Ax - \frac{\hbar^2 x^2}{2mb^4}\right) \phi_0 \; dx = \frac{2Ab}{\sqrt{\pi}} - \frac{3\hbar^2}{4mb^2} .$$

One possible choice for b is such that the first-order energy contribution is zero i.e.

$$b^3 = \frac{3\hbar^2\sqrt{\pi}}{8Am} .$$

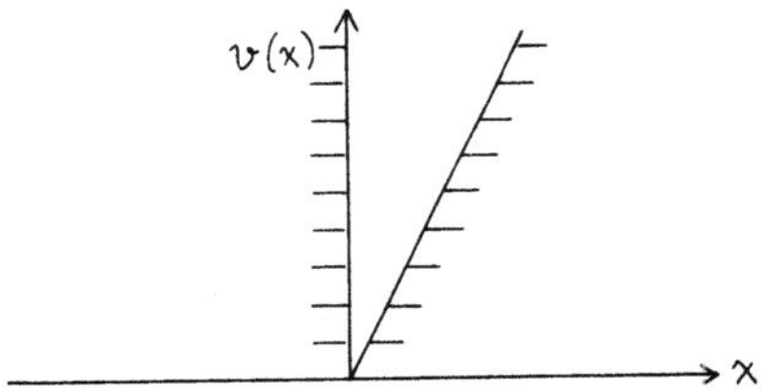

Figure 11.2. Potential in Example 11.2.

with this choice of

$$b, E_0 = \left(\frac{24}{\pi}\right)^{1/3}\left(\frac{\hbar^2 A^2}{m}\right)^{1/3} \doteq 1.97 \left(\frac{\hbar^2 A^2}{m}\right)^{1/3}$$

to first order.

The second-order term is

$$\sum_{n=1,\,2\ldots} \frac{<0|Ax - \frac{\hbar^2 x^2}{2mb^4}|n><n|Ax - \frac{\hbar^2 x^2}{2mb^4}|0>}{-\frac{\hbar^2}{mb^2}\,2n} , \tag{11.16}$$

since

$$\varepsilon_n = (2n + \tfrac{3}{2})\frac{\hbar^2}{mb^2} .$$

One can evaluate this expression using

$$\phi_1 = |1\rangle = \sqrt{\frac{2}{3}}\left(\frac{1}{b\sqrt{\pi}}\right)^{1/2} e^{-x^2/2b^2}\left(2\frac{x^3}{b^3} - 3\frac{x}{b}\right)$$

etc.

Considering just the n = 1 term in expression (11.16) one has

$$\langle 0|Ax - \frac{\hbar^2 x^2}{2mb^4}|1\rangle = \sqrt{\frac{2}{3}}\frac{Ab}{\sqrt{\pi}} - \sqrt{\frac{3}{2}}\frac{\hbar^2}{2mb^2} = -\frac{Ab}{\sqrt{\pi}}\sqrt{\frac{2}{3}}$$

with the above choice of b. Thus Equation (11.16) becomes

$$\frac{b^2A^2\frac{2}{3\pi}}{\frac{\hbar^2}{mb^2}2} = -\frac{1}{8}\left(\frac{3}{8\pi}\right)^{1/3}\left(\frac{\hbar^2A^2}{m}\right)^{1/3} = -0.06\left(\frac{\hbar^2A^2}{m}\right)^{1/3}. \tag{11.17}$$

All additional second-order terms are also negative hence add to the magnitude of the second-order result, but should be small compared to Equation (11.17). Indeed the next term in second order (n = 2) is:

$$\frac{|\langle 0|Ax - \frac{\hbar^2x^2}{2mb^4}|2\rangle|^2}{-\frac{4\hbar^2}{mb^2}} = \frac{|-\frac{bA}{\sqrt{30\pi}} - 0|^2}{-\frac{4\hbar^2}{mb^2}} = -\frac{b^4mA^2}{120\pi\hbar^2} =$$

$$-\frac{1}{320}\left(\frac{3}{8\pi}\right)^{1/3}\left(\frac{A^2\hbar^2}{m}\right)^{1/3} = -0.0015\left(\frac{A^2\hbar^2}{m}\right)^{1/3},$$

which is considerably smaller than Equation (11.17).

Hence to second order

$$E_0 = 1.91\left(\frac{\hbar^2A^2}{m}\right)^{1/3}$$

which can be compared with the variational approach result for this problem, (Example 10.2)

$$E_0 \leq 1.86\left(\frac{A^2\hbar^2}{m}\right)^{1/3}.$$

The third-order term is

$$\sum_{n,\,p\neq 0} \frac{\langle 0|Ax - \frac{\hbar^2 x^2}{2mb^4}|n\rangle\langle n|Ax - \frac{\hbar^2 x^2}{2mb^4}|p\rangle\langle p|Ax - \frac{\hbar^2 x^2}{2mb^4}|0\rangle}{\left(-2n\frac{\hbar^2}{mb^2}\right)\left(-2p\frac{\hbar^2}{mb^2}\right)}, \quad (11.18)$$

since the choice of b earlier in this problem makes the second third-order term zero (see Equation (11.14)).

Taking only the term $|n\rangle = |m\rangle = |1\rangle$ in Equation (11.18) one obtains an approximate third order result:

$$\frac{|\langle 0|Ax - \frac{\hbar^2 x^2}{2mb^4}|1\rangle|^2\langle 1|Ax - \frac{\hbar^2 x^2}{2mb^4}|1\rangle}{\frac{4\hbar^4}{m^2 b^4}} = \frac{\frac{b^2A^2}{\pi}\frac{2}{3}}{\frac{4\hbar^4}{m^2b^4}}\left(\frac{3bA}{\sqrt{\pi}} - \frac{7\hbar^2}{4mb^2}\right)$$

$$= -\frac{5}{2^8}\left(\frac{3}{\pi}\right)^{1/3}\left(\frac{A^2\hbar^2}{m}\right)^{1/3} = -0.02\left(\frac{A^2\hbar^2}{m}\right)^{1/3}$$

which is smaller than the second-order correction.

Thus to this approximation $E_0 \sim 1.89(A^2\hbar^2/m)^{1/3}$.

If one uses instead $b^3 = (3\sqrt{\pi}\,\hbar^2)/4mA$ which minimizes the energy to first order (see Example 10.2), one obtains $E_0 = 1.86\,(\hbar^2A^2/m)^{1/3}$ in first order and no contribution in second and third order from the state $n = 1$.

The exact result (valid for large arguments of the relevant Airy function (cf. Expression (3.18)) is:

$$E = \left(\frac{3}{4}\right)^{2/3}\left(\frac{A^2\hbar^2}{m}\right)^{1/3}\left(\frac{3\pi}{2\sqrt{2}}\right)^{2/3} = \left(\frac{9\pi}{8\sqrt{2}}\right)^{2/3}\left(\frac{A^2\hbar^2}{m}\right)^{1/3} \doteq 1.84\left(\frac{A^2\hbar^2}{m}\right)^{1/3}.$$

With the first choice of

$$b, \left\{b_1 = \left(\frac{3\hbar^2\sqrt{\pi}}{8Am}\right)^{1/3} \quad \text{which makes } V_{00} = 0\right\},$$

to second order in perturbation theory the exact ground state wave-function becomes:

$$\psi_0 = \phi_0(b_1) + \frac{1}{8}\sqrt{\frac{3}{2}}\,\phi_1(b_1) + \frac{5}{2^7}\sqrt{\frac{3}{2}}\,\phi_1(b_1) \sim \phi_0(b_1) +$$

$$+ (0.153+0.048)\phi_1(b_1) = \phi_0(b_1) + 0.201\phi_1(b_1) \quad (11.19)$$

if one includes only one excited state in expression (11.15).

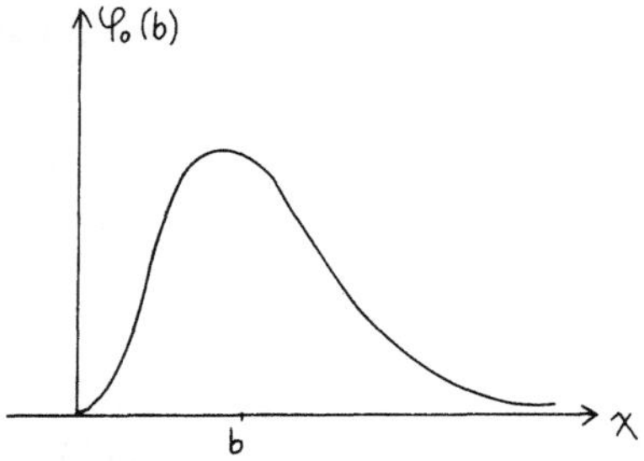

Figure 11.3. $\phi_0(b)$ in Example 11.2 vs x.

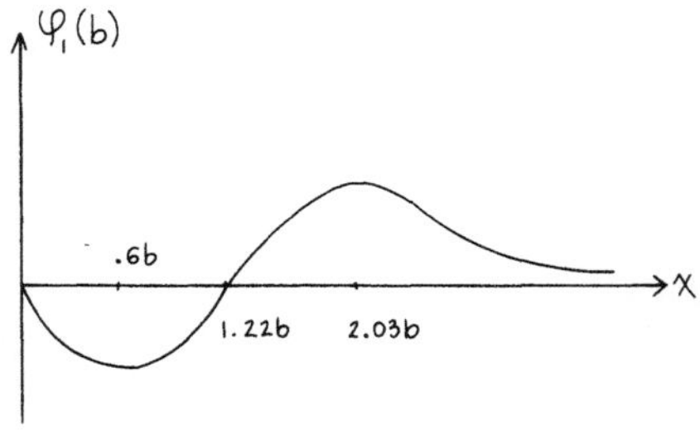

Figure 11.4. $\phi_1(b)$ in Example 11.2 vs x.

With the second choice of

$$b, \left\{b_2 = \left(\frac{3\hbar^2\sqrt{\pi}}{4Am}\right)^{1/3}\right. ,$$

which minimizes the energy to first order, i.e. makes $\left.\partial H_{00}/\partial b = 0\right\}$, the matrix element V_{01} is zero so there is no contribution to the exact ground-state wavefunction from ϕ_1 for this choice of b:

$$\psi_0 \simeq \phi_0(b_2) \text{ to second order in V,} \tag{11.20}$$

Expressions (11.19) and (11.20) are less dissimilar than they look since they involve different b's with $b_1 \sim 0.794b_2$. Thus though $\phi_0(b_1)$ of Equation (11.19) is more compressed than $\phi_0(b_2)$ of Equation (11.20), this is compensated for by the small admixture in the former expression of $\phi_1(b_1)$ as can be seen by considering Figures 11.3 and 11.4.

EXAMPLE 11.3. Consider

$$\begin{aligned} H &= T + \frac{1}{2} m\omega^2\rho^2 \qquad 0 < \rho \le a \quad \text{where } \rho = \sqrt{x^2 + y^2} \\ &= T \qquad \rho > a. \end{aligned} \tag{11.21}$$

To find the ground state energy of this system using perturbation theory one can write

$$H = T + \frac{1}{2} m\omega^2\rho^2 + V(\rho) = H_0 + V(\rho),$$

where

$$V(\rho) = 0 \quad \rho < a, \quad V(\rho) = -\frac{1}{2} m\omega^2\rho^2 \quad \rho > a.$$

The ground state wavefunction for H_0 is (see Equation (8.14))

$$w_{00}(\rho) = \left(\frac{2}{b^2}\right)^{1/2} \rho^{\frac{1}{2}} e^{-\rho^2/2b^2}, \quad \text{with} \quad b = \sqrt{\frac{\hbar}{m\omega}},$$

$$\varepsilon_0 = \hbar\omega .$$

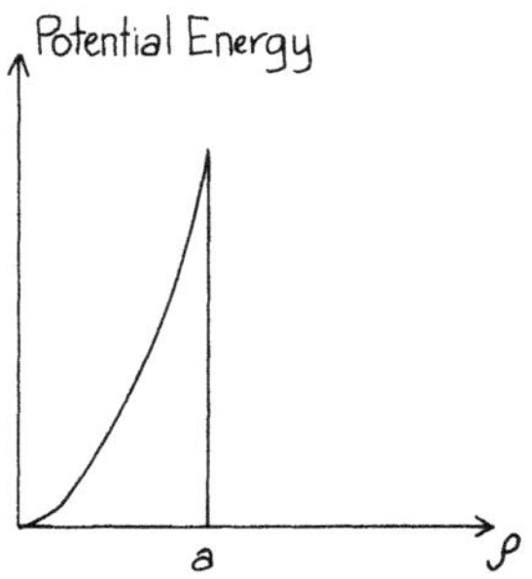

Figure 11.5. Potential in Example 11.3.

To first order in V one then has

$$E_0 = \hbar\omega - \frac{2}{b^2}\int_a^{\infty} \rho e^{-\rho^2/b^2}\left(\frac{1}{2} m\omega^2\rho^2\right)d\rho =$$

$$= \hbar\omega - \frac{\hbar\omega}{2}\left\{- e^{-\rho^2/b^2}\left(1 + \frac{\rho^2}{b^2}\right)\Big|_a^{\infty}\right\} =$$

$$= \hbar\omega - \frac{\hbar\omega}{2}\left(1 + \frac{m\omega a^2}{\hbar}\right) e^{-\frac{m\omega a^2}{\hbar}}, \tag{11.22}$$

which is expected to be accurate if $\hbar\omega << m\omega^2 a^2$.

To second order the first state which contributes to the energy is $\omega_{10}(\rho)$ at energy $3\hbar\omega$

$$= \frac{\left(\int_a^\infty \omega_{10}(\rho)\left(-\frac{1}{2}\, m\omega^2\rho^2\right)\, \omega_{00}(\rho)\, d\rho\right)^2}{-2\hbar\omega} =$$

$$= \frac{\left(-\frac{1}{2}\,\hbar\omega\left\{e^{-\rho^2/b^2}\left(1 + \frac{\rho^2}{b^2} + \frac{\rho^4}{b^4}\right)\Big|_a^\infty\right\}\right)^2}{-2\hbar\omega}$$

$$= -\frac{\hbar\omega}{8}\left(e^{-\frac{m\omega a^2}{\hbar}}\left(1 + \frac{m\omega a^2}{\hbar} + \frac{m^2\omega^2a^4}{\hbar^2}\right)\right)^2 = -\frac{\hbar\omega}{8}\, e^{-\frac{2m\omega a^2}{\hbar}}\left(1+\frac{m\omega a^2}{\hbar} + \frac{m^2\omega^2a^4}{\hbar^2}\right)^2.$$

Hence approximately

$$E_0 = \hbar\omega - \frac{\hbar\omega}{2}\left(1+ \frac{m\omega a^2}{\hbar}\right)e^{-\frac{m\omega a^2}{\hbar}} - \frac{\hbar\omega}{8}\, e^{-\frac{2m\omega a^2}{\hbar}}\left(1+ \frac{m\omega a^2}{\hbar} + \frac{m^2\omega^2a^4}{\hbar^2}\right)^2. \tag{11.23}$$

Obviously this analysis is incorrect if $E > \frac{1}{2}\, m\omega^2a^2$ for reasons similar to those mentioned in Example 11.1.

EXAMPLE 11.4. Consider

$$H = T + A|x|^p \quad \text{(all } x\text{)}. \tag{11.24}$$

This may be written

$$H = T + \frac{\hbar^2}{2mb^4}\, x^2 + \left\{A|x|^p - \frac{\hbar^2}{2mb^4}\, x^2\right\} \text{ and } A|x|^p - \frac{\hbar^2}{2mb^4}\, x^2$$

can be treated as a perturbation. The ground state energy of this system can be immediately calculated using first-order perturbation theory

$$\varepsilon_0 = \frac{\hbar^2}{2mb^2} \quad \text{and} \quad \phi_0(x) = \left(\frac{1}{b\sqrt{\pi}}\right)^{1/2} e^{-x^2/2b^2}.$$

If p is odd one has:

$$\langle 0|A|x|^p - \frac{\hbar^2}{2mb^4}\, x^2|0\rangle = \frac{Ab^p}{\sqrt{\pi}}\left(\frac{p-1}{2}\right)! - \frac{\hbar^2}{4mb^2}\,.$$

The choice

$$b^{p+2} = \frac{\hbar^2\sqrt{\pi}}{4mA\left(\frac{p-1}{2}\right)!}$$

makes the first-order energy zero and yields

$$E_0 \simeq \frac{2^{\frac{2-p}{2+p}}}{\pi^{\frac{1}{p+2}}}\left(\left(\frac{p-1}{2}\right)!\,\frac{\hbar^p A}{m^{p/2}}\right)^{\frac{2}{p+2}} . \tag{11.25}$$

If p is even one has

$$<0|Ax^p - \tfrac{1}{2}m\omega^2x^2|0> = \frac{Ab^p(p-1)!!}{2^{p/2}} - \frac{\hbar^2}{4mb^2} .$$

The choice

$$b^{p+2} = \frac{\hbar^2 2^{p/2-2}}{mA(p-1)!!}$$

makes the first-order energy zero and yields:

$$E_0 \simeq \frac{1}{2^{\frac{2p-2}{p+2}}}\left(\left(p-1\right)!!\,\frac{\hbar^p A}{m^{p/2}}\right)^{\frac{2}{p+2}} . \tag{11.26}$$

These results may be combined:

$$E_0 = \frac{1}{2^{\frac{p-2}{p+2}}}\left(\frac{\Gamma(\frac{p+1}{2})\ \hbar^p A}{\pi^{1/2}\ m^{p/2}}\right)^{\frac{2}{p+2}} \text{ for arbitrary p.} \tag{11.27}$$

These may be compared with a qualitatively similar result namely Equation (5.2). If instead one chooses b so the ground state energy will be a minimum to first order (cf. Ch. 10) i.e.

$$<0|H_0+V|0> = <0|T + A|x|^p|0> = <0|H|0>$$

is a minimum, for odd p this implies

$$b^{p+2} = \frac{\hbar^2\sqrt{\pi}}{2mAp\left(\frac{p-1}{2}\right)!}$$

while for even p

$$b^{p+2} = \frac{\hbar^2 2^{\frac{p}{2}-1}}{mAp(p-1)!!} .$$

For p odd in this case

$$E_0 \doteq \frac{p+2}{\pi^{\frac{1}{p+2}}\ 2^{\frac{2p+2}{p+2}}}\left(\frac{A\hbar^p\left(\frac{p-1}{2}\right)!}{m^{p/2}\ p^{p/2}}\right)^{2/(p+2)}$$

while for p even

$$E_0 = \frac{p+2}{2^{\frac{3p+2}{p+2}}} \left(\frac{A\hbar^p (p-1)!!}{m^{p/2}\ p^{p/2}}\right)^{2/(p+2)} .$$

These results may be combined yielding

$$E_0 \simeq \frac{p+2}{2^{\frac{2p+2}{p+2}}} \left(\frac{A\hbar^p\ \Gamma\left(\frac{p+1}{2}\right)}{m^{p/2} p^{p/2} \pi^{1/2}}\right)^{2/(p+2)} \tag{11.28}$$

for arbitrary p. For p = 2 Equations (11.27) and (11.28) both yield $\hbar\omega/2$ which as expected is the right result (where $A = m\omega^2/2$).

EXAMPLE 11.5. Consider $H = H_0 + V(x, y)$ where

$$H_0 = \frac{p_x^2}{2m} + \frac{p_y^2}{2m} + \frac{1}{2} m\omega^2 (x^2+y^2)$$

while

$$V(x, y) = \lambda m\omega^2 xy.$$

Suppose one wishes to find the ground state energy (non-degenerate) of this system.

The unperturbed energy of this system is $\hbar\omega$ with corresponding wavefunction $\phi_0(x)\ \phi_0(y)$ (cf. Expression (12.4)).

The first-order correction to this energy is zero since

$$\lambda m\omega^2 \iint \phi_0^*(x)\ \phi_0^*(y)\ xy\phi_0(x)\ \phi_0(y)\ dx\ dy = 0.$$

The only term which contributes in second order because of the nature of this particular V is $\phi_1(x)\ \phi_1(y)$ at energy $3\hbar\omega$.

Hence the second order contribution to the energy is

$$\frac{(\lambda m\omega^2)^2 |\langle\phi_0(x)\phi_0(y)xy\phi_1(x)\phi_1(y)\rangle|^2}{-2\hbar\omega} =$$

$$- \frac{\lambda^2 4}{2\hbar\omega} |\langle\phi_0|\frac{m\omega^2 x^2}{2}|\phi_0\rangle|^2 = - \frac{\lambda^2\hbar\omega}{8}$$

in agreement with the exact result (see Example (12.9)),

$$E_0 = \frac{1}{2}\hbar\omega \left\{\sqrt{1-\lambda} + \sqrt{1+\lambda}\right\} = \hbar\omega - \frac{\lambda^2\hbar\omega}{8} + \ldots \quad . \tag{11.29}$$

The exact normalized wavefunction for this state (Example (12.9)) is $\psi_0 = \phi_0(X, \omega = \omega_1)\phi_0(Y, \omega = \omega_2)$, where

$$X = \frac{1}{\sqrt{2}}(x - y) \quad \text{and} \quad \omega_1 = \omega\sqrt{1 - \lambda}$$

$$Y = \frac{1}{\sqrt{2}}(x + y) \qquad\qquad \omega_2 = \omega\sqrt{1 + \lambda}$$

while

$$\phi_0(\xi, \omega) = \left(\frac{m\omega}{\hbar\pi}\right)^{1/4} e^{-\frac{m\omega\xi^2}{2\hbar}}$$

i.e.

$$\psi_0(x, y) = \left(\frac{m\omega}{\hbar\pi}\right)^{1/2} (1-\lambda^2)^{1/8}\, e^{-\frac{m\omega\sqrt{1-\lambda}\, X^2}{2\hbar}}\, e^{-\frac{m\omega\sqrt{1+\lambda}\, Y^2}{2\hbar}} ,$$

or

$$\psi_0(x, y) = \left(\frac{m\omega}{\hbar\pi}\right)^{1/2}\left(1-\lambda^2\right)^{1/8} e^{-\frac{m\omega x^2}{2\hbar}(\sqrt{1-\lambda} + \sqrt{1+\lambda})/2}\, e^{-\frac{m\omega}{2\hbar} y^2(\sqrt{1-\lambda} + \sqrt{1+\lambda})/2} \times$$

$$e^{-\frac{m\omega xy}{2\hbar}\{\sqrt{1+\lambda} - \sqrt{1-\lambda}\}} . \tag{11.30}$$

If $\lambda \ll 1$ one can write Equation (11.30) as

$$\psi_0(x, y) \simeq \left(1 - \frac{\lambda^2}{8}\right)\phi_0(x, \omega)\phi_0(y, \omega)\, e^{\frac{m\omega}{2\hbar 8}\lambda^2(x^2+y^2)}\, e^{-\frac{m\omega xy\lambda}{2\hbar}} .$$

$$\simeq \phi_0(x)\phi_0(y)\left(1 - \frac{\lambda m\omega xy}{2\hbar} + \frac{\lambda^2}{8}\left\{-1+ \frac{m\omega}{2\hbar}(x^2+y^2)+\frac{m^2\omega^2}{\hbar^2} x^2y^2\right\}+\dots\right). \tag{11.31}$$

Perturbation theory on the other hand yields, to second order, for the wavefunction:

$$\psi_0 \sim N\Big(\phi_0(x)\phi_0(y) - \frac{\lambda}{4}\phi_1(x)\phi_1(y) + \frac{\lambda^2}{8\sqrt{2}}\phi_2(x)\phi_0(y) + \frac{\lambda^2}{8\sqrt{2}}\phi_0(x)\phi_2(y) +$$

$$+ \frac{\lambda^2}{16}\phi_2(x)\phi_2(y)\Big) , \tag{11.32}$$

since

$$(\phi_1(x)\phi_1(y)|V|\phi_1(x)\phi_1(y)) = 0$$

$$(\phi_1(x)\phi_1(y)|V|\phi_0(x)\phi_0(y)) = \frac{\lambda\hbar\omega}{2} ,$$

$$(\phi_2(x)\phi_0(y)|V|\phi_1(x)\phi_1(y)) = \frac{\lambda\hbar\omega}{\sqrt{2}} ,$$

$$(\phi_2(x)\phi_2(y)|V|\phi_1(x)\phi_1(y)) = \lambda\hbar\omega$$

and where N = 1 if one assumes ψ_0 is a cross normalized function i.e. $(\phi_0(x)\phi_0(y)|\psi_0(x, y)) = 1$, while it equals $\sim (1 + \lambda^2/16 + \ldots)^{-1/2}$ if one assumes (as in the case of Equation (11.30)) that $(\psi_0(x, y)|\psi_0(x, y)) = 1$.

Expressions (11.31) and (11.32) are identical to order λ^2 if one substitutes $N \sim 1 - \lambda^2/32$ in Equation (11.32), since

$$\phi_1(x) = \sqrt{\frac{2m\omega}{\hbar}}\, x\, \phi_0(x)$$

and

$$\phi_2(x) = \frac{1}{\sqrt{2}}\left(-1 + \frac{2m\omega}{\hbar}x^2\right)\phi_0(x) .$$

EXAMPLE 11.6. Consider

$$H = \frac{p^2}{2m} + V_0\delta(x) \quad -a < x < a ,$$

whereas if $|x| > a$, $V = \infty$.

To obtain the exact result one writes the Schrödinger equation as follows:

$$(H_0 - E)\psi(x) = -V_0\delta(x)\psi(x),$$

where $H_0 = \frac{p^2}{2m}$, $V = V_0\delta(x)$ and $x \le |a|$.

Thus

$$\psi(x) = -\frac{V_0\delta(x)}{\frac{p^2}{2m} - E}\psi(x) . \tag{11.33}$$

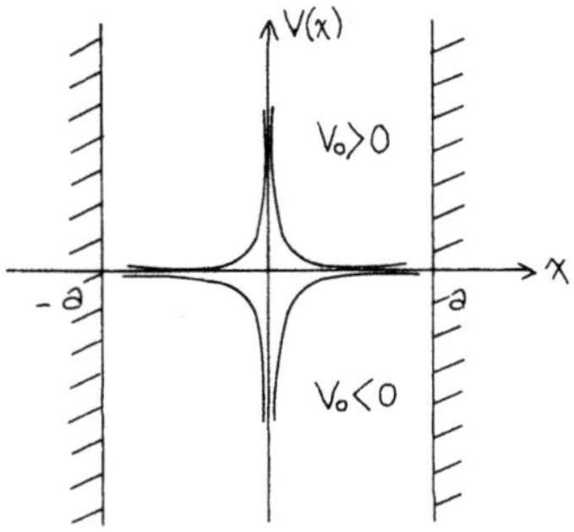

Figure 11.6. Potential Energy in Example 11.6.

Using closure (cf. Equation (2.6)) one may insert the set $|n>$ of eigenfunctions of the potential $V = 0 \; |x| < a$, $V = \infty |x|>a$ between $1/\left(\frac{p^2}{2m} - E\right)$ and $V_0\delta(x)$ in Equation (11.33) {the odd eigenfunctions do not contribute to $(n|V_0\delta(x)|\psi(x))$}. One then obtains:

$$\psi(x) = - \sum_{n=1,\,3\ldots} \frac{|n)(n|V_0\delta(x)|\psi(x))}{\frac{n^2h^2}{32ma^2} - E} =$$

$$= V_0\psi(0) \sum_{n=1,\,3\ldots} \frac{\phi_n(x)\phi_n^*(0)}{E - \frac{n^2h^2}{32ma^2}} \quad . \tag{11.34}$$

Substituting the value $x = 0$ in expression (11.34) one obtains

$$\psi(0) = - \sum_{n=1,\,3\ldots} \frac{\phi_n^*(0)\phi_n(0)V_0\psi(0)}{\frac{n^2h^2}{32ma^2} - E} \quad ,$$

i.e.

$$1 = - \frac{V_0}{a} \sum_{n=1,\,3\ldots} \frac{1}{\frac{n^2h^2}{32ma^2} - E}$$

since

$$\phi_n(x) = \frac{1}{\sqrt{a}} \cos \frac{n\pi x}{2a} \quad ,$$

$$\varepsilon_n = \frac{n^2h^2}{32ma^2} \quad , \; n=1,\, 3,\, 5\, \ldots$$

i.e.

$$\frac{h^2}{32V_0ma} = \sum_{n_{odd}} \frac{1}{\beta - n^2} \tag{11.35}$$

where

$$\beta = \frac{32ma^2E}{h^2} \quad .$$

If one plots the two sides of this equation one obtains Figure 11.7, where the intersections correspond to acceptable values of E.

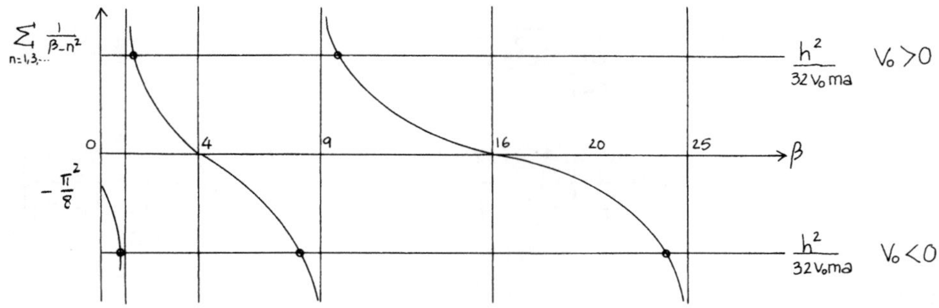

Figure 11.7. Exact energies in Example 11.6, graphical solution of Equation (11.35)

If $\beta = q^2 + \delta$

$$\frac{h^2}{32V_0ma} \doteq \frac{1}{\delta} , \quad \delta \simeq \frac{32V_0ma}{h^2}$$

i.e.

$$\beta \simeq q^2 + \frac{32V_0ma}{h^2} ,$$

or

$$E = \frac{h^2q^2}{32ma^2} + \frac{V_0}{a} = \varepsilon_q + \frac{V_0}{a} .$$

A more accurate calculation yields:

$$\frac{h^2}{32V_0ma} = \frac{1}{\delta} + \sum_{\substack{n\neq q\\(\text{odd})}} \frac{1}{q^2-n^2+\delta} \simeq \frac{1}{\delta} + \sum_{\substack{n\neq q\\(\text{odd})}} \frac{1}{q^2-n^2} - \delta \sum_{\substack{n\neq q\\(\text{odd})}} \frac{1}{(q^2-n^2)^2} ,$$

i.e.

$$c_1 \delta^2 + c_2\delta - 1 = 0$$

implying

$$\delta \sim \frac{1}{c_2} - \frac{c_1}{c_2^3}$$

where

$$c_1 = \sum_{\substack{n\neq q\\(\text{odd})}} \frac{1}{(q^2-n^2)^2} ,$$

$$c_2 = \frac{h^2}{32V_0ma} - \sum_{\substack{n\neq q\\(\text{odd})}} \frac{1}{q^2-n^2} .$$

Thus

$$\delta = \frac{32V_0ma}{h^2}\left\{1 - \sum_{\substack{n\neq q\\(\text{odd})}} \frac{V_0/a}{\varepsilon_q-\varepsilon_n}\right\}^{-1} - \frac{32V_0ma}{h^2} \sum_{\substack{n\neq q\\(\text{odd})}} \frac{(V_0/a)^2}{(\varepsilon_q-\varepsilon_n)^2}\left\{1- \sum_{\substack{n\neq q\\(\text{odd})}} \frac{V_0/a}{\varepsilon_q-\varepsilon_n}\right\}^{-3} .$$

Since

$$E = \frac{h^2}{32ma^2}(q^2 + \delta),$$

$$E = \varepsilon_q - \frac{V_0}{a} + \frac{V_0^2}{a^2}\sum_{\substack{n\neq q\\\text{odd}}} \frac{1}{\varepsilon_q-\varepsilon_n} + \frac{V_0^3}{a^3}\sum_{\substack{n\\m\}\neq q\\\text{odd}}} \frac{1}{(\varepsilon_q-\varepsilon_n)}\frac{1}{(\varepsilon_q-\varepsilon_m)} - \frac{V_0^3}{a^3}\sum_{\substack{n\neq q\\\text{odd}}} \frac{1}{(\varepsilon_q-\varepsilon_n)^2}, \tag{11.36}$$

which is just the perturbation result for E_q (q = odd) to third order since if r, s are odd $(\phi_r|V|\phi_s) = V_0/a$ for this particular V and set of eigenfunctions of H_0. If q = even there is no correction to the unperturbed energy ε_q from this particular perturbation.

As concerns the exact wavefunction, from Equation (11.34):

$$\psi(x) = \frac{V_0\psi(0)}{\sqrt{a}} \frac{\phi_q(x)}{E-\frac{q^2h^2}{32ma^2}} + \frac{V_0\psi(0)}{\sqrt{a}} \sum_{\substack{n\neq q\\\text{odd}}} \frac{\phi_n(x)}{E-\frac{n^2h^2}{32ma^2}} .$$

But if $(\phi_q|\psi(x)) = 1$, this implies

$$\frac{V_0\psi(0)}{\sqrt{a}} \frac{1}{E-\frac{q^2h^2}{32ma^2}} = 1.$$

With this constraint

$$\psi(x) = \phi_q(x) + \left(E - \frac{q^2h^2}{32ma^2}\right) \sum_{\substack{n\neq q\\(\text{odd})}} \frac{\phi_n(x)}{E-\frac{n^2h^2}{32ma^2}} . \tag{11.37}$$

Substituting Equation (11.36) into Equation (11.37),

$$\psi(x) = \phi_q(x) + \frac{V_0}{a}\left\{1 + \frac{V_0}{a}\sum_{\substack{m\neq q\\ \text{odd}}}\frac{1}{\varepsilon_q-\varepsilon_m} + \dots\right\}\sum_{\substack{n\neq q\\ \text{odd}}}\frac{\phi_n(x)}{(\varepsilon_q-\varepsilon_n)\left\{1+\frac{V_0}{a}\frac{1}{\varepsilon_q-\varepsilon_n}+\dots\right\}}$$

i.e.

$$\psi(x) = \phi_q(x) + \frac{V_0}{a}\sum_{\substack{n\neq q\\ \text{odd}}}\frac{\phi_n(x)}{\varepsilon_q-\varepsilon_n} + \left(\frac{V_0}{a}\right)^2\sum_{\substack{n\neq q\\ \text{odd}}}\frac{\phi_n(x)}{(\varepsilon_q-\varepsilon_m)(\varepsilon_q-\varepsilon_n)} - \left(\frac{V_0}{a}\right)^2\sum_{\substack{n\neq q\\ \text{odd}}}\frac{\phi_n}{(\varepsilon_q-\varepsilon_n)^2}+\dots, \tag{11.38}$$

which is just the perturbation result for ψ_q to second order in V.

If q is even

$$\psi_q(x) = \phi_q(x) \quad \left(\text{where } \phi_q(x) = \sqrt{\frac{1}{a}}\sin\frac{q\pi x}{2a}, \quad q = 2, 4, \dots\right),$$

i.e. the perturbation affects neither the unperturbed energy nor the unperturbed wavefunctions of the system in this case. The reason for this is because $\phi_q(0) = 0$, i.e. the system does not feel the presence of V when in these states.

Explicitly the wavefunction described by Equation (11.38) is to first order:

$$\psi(x) = \frac{1}{\sqrt{a}}\cos\frac{q\pi x}{2a} + \frac{V_0}{a}\sum_{n\neq q}\frac{\frac{1}{\sqrt{a}}\cos\frac{n\pi x}{2a}}{(q^2-n^2)\frac{h^2}{32ma^2}} \qquad q,\ n \text{ odd.} \tag{11.39}$$

For this particular Hamiltonian one can also obtain the <u>exact</u> eigenfunctions of this system in a form which unlike Equation (11.37) does not involve infinite sums. This can be done by writing

$$\psi(x) = A\sin k(a - x) \qquad 0 < x < a,$$

$$\psi(x) = B\sin k(a + x) \qquad -a < x < 0.$$

This combination of eigenfunctions of H_0 satisfies the requirement that $\psi(a) = \psi(-a) = 0$.

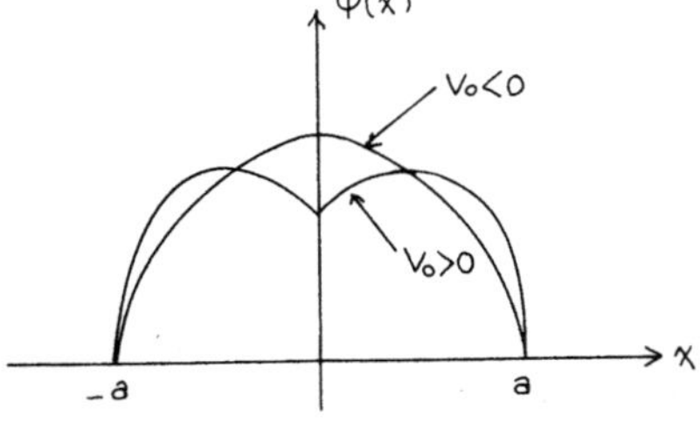

Figure 11.8. Two possible even eigenfunctions of Example 11.6.

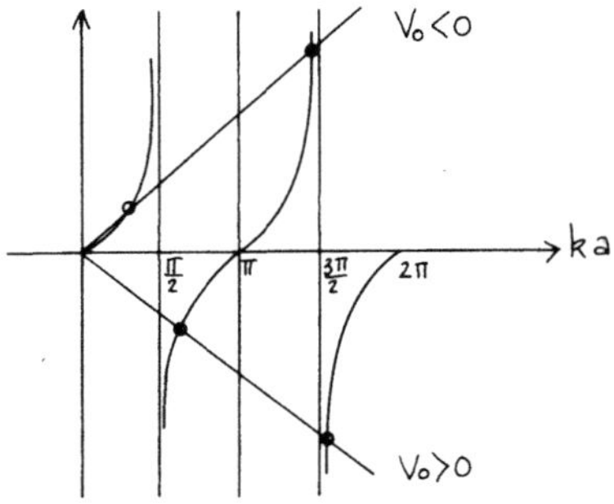

Figure 11.9. Plot of transcendental Equation (11.40).

Additionally continuity of $\psi(x)$ at $x = 0$ implies A sinka = B sinka i.e. A = B. However, the derivative of ψ has a discontinuity at $x = 0$.

$$\left.\frac{d\psi}{dx}\right|_{x=\varepsilon} - \left.\frac{d\psi}{dx}\right|_{x=-\varepsilon} = \frac{2mV_0}{\hbar^2}\,\psi(0) \quad \text{(cf. Equation (6.10)).}$$

This implies

$$- A\,k\,\cos ka - A\,k\,\cos ka = \frac{2mV_0}{\hbar^2}\,A\sin ka,$$

or

$$\tan ka = -\frac{\hbar^2 k}{mV_0}\,. \tag{11.40}$$

Thus the even parity solutions $\psi(x)$ are given by

$$\psi(x) = A\sin k\,(a - |x|) \qquad -a < x < a,$$

where k is given by Equation (11.40).

Normalization of $\psi(x)$ additionally requires

$$|A|^2 \int_{-a}^{a} \sin^2 k(a-|x|)\,dx = 1,$$

or

$$\psi(x) = \frac{\sin k\,(a-|x|)}{\sqrt{a\left(1-\frac{\sin 2ka}{2ka}\right)}} \tag{11.41}$$

which is illustrated in Figure 11.8.

Plotting tan ka and $-\frac{\hbar^2 ka}{mV_0 a}$ of Equation (11.40) vs ka (see Figure 11.9) shows $ka = \frac{q\pi}{2} + \delta$ where δ is small (q = odd).

Thus $\tan ka = -\cot\delta = -\frac{\hbar^2 k}{mV_0}$ or $\tan\delta \sim \delta = \frac{mV_0}{\hbar^2 k}$,

and

$$ka \simeq \frac{q\pi}{2} + \frac{mV_0 a}{\hbar^2 ka} \simeq \frac{q\pi}{2} + \frac{2mV_0 a}{\hbar^2 q\pi} .$$

Hence one can expand the exact wavefunctions Equation (11.41) in a Taylor series about $k_0 = q\pi/2a$ where $k - k_0 = (2mV_0)/\hbar^2 q\pi)$ and q is odd.

Thus

$$\psi(x) \simeq \sin\frac{q\pi}{2}\left[\frac{1}{\sqrt{a}}\cos\frac{q\pi x}{2a} + \frac{\frac{V_0}{a}\left((1-\frac{x}{a})\frac{q\pi}{4\sqrt{a}}\sin\frac{q\pi x}{2a} - \frac{1}{4\sqrt{a}}\cos\frac{q\pi x}{2a}\right)}{\frac{q^2\hbar^2}{32ma^2}}\right] \tag{11.42}$$

as opposed to Equation (11.39) (to within an unimportant overall phase).

Hence, by comparing the terms in V_0/a of Equations (11.39) and (11.42) one obtains:

$$\left(1 - \frac{x}{a}\right)\frac{q\pi}{4}\sin\frac{q\pi x}{2a} - \frac{1}{4}\cos\frac{q\pi x}{2a} = \sum_{n\neq q}\frac{\cos\frac{n\pi x}{2a}}{1-\left(\frac{n}{q}\right)^2} ,$$

or

$$\frac{q}{2}\left(\frac{\pi}{2} - u\right)\sin qu - \frac{1}{4}\cos qu = \sum_{n\neq q}\frac{\cos nu}{1 - \left(\frac{n}{q}\right)^2} \tag{11.43}$$

where $u = \frac{\pi x}{2a}$

EXAMPLE 11.7. Consider the Hamiltonian:

$$H = \frac{p^2}{2m} + \frac{1}{2}m\omega^2 x^2 + V_0\, e^{-\frac{x^2}{\varepsilon}}$$

$$= H_0 + V$$

where $H_0 = T + \frac{1}{2}m\omega^2 x^2$ and $V = V_0\, e^{-x^2/\varepsilon}$.

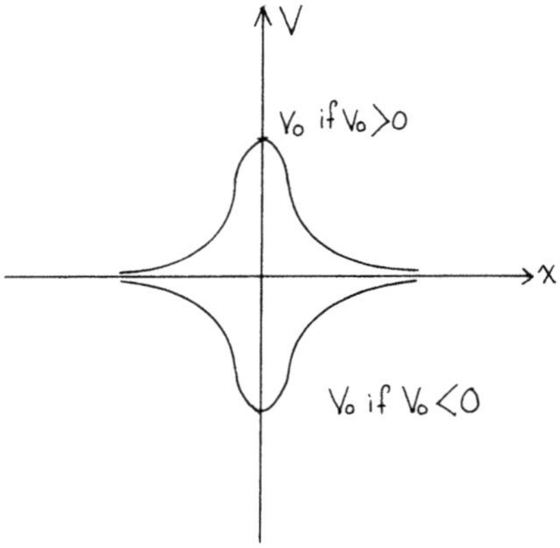

Figure 11.10. Perturbation V in Example 11.7.

To first order the ground state energy of this system is:

$$E_0 = \frac{\hbar\omega}{2} + \frac{V_0}{b\sqrt{\pi}} \int_{-\infty}^{+\infty} e^{-x^2/b^2} e^{-x^2/\varepsilon} dx \quad \text{since} \quad \phi_0 = \left(\frac{1}{b\sqrt{\pi}}\right)^{\frac{1}{2}} e^{-x^2/2b^2},$$

where $b = \sqrt{\frac{\hbar}{m\omega}}$.

$$E_0 = \frac{\hbar\omega}{2} + \frac{V_0}{b\sqrt{\pi}\left\{\frac{1}{b^2} + \frac{1}{\varepsilon}\right\}^{\frac{1}{2}}} \int_{-\infty}^{+\infty} e^{-x^2\left\{\frac{1}{b^2} + \frac{1}{\varepsilon}\right\}} dx\left\{\frac{1}{b^2} + \frac{1}{\varepsilon}\right\}^{\frac{1}{2}},$$

i.e.

$$E_0 = \frac{\hbar\omega}{2} + \frac{V_0}{\left\{1 + \frac{b^2}{\varepsilon}\right\}^{\frac{1}{2}}}. \tag{11.44}$$

One notes this reduces to the correction to the unperturbed energy introduced by a delta function potential if $V_0 \to \sqrt{\frac{1}{\pi\varepsilon}}$ and $\varepsilon \to 0$ since one representation of the delta function is

$$\delta(x) = \frac{1}{\sqrt{\pi}} \lim_{\varepsilon\to 0} \sqrt{\frac{1}{\varepsilon}}\, e^{-x^2/\varepsilon} \qquad \text{(see Equation (2.15)).}$$

EXAMPLE 11.8. Consider the Hamiltonian

$$H = \frac{p^2}{2m} + \frac{1}{2} m\omega^2 x^2 + \frac{\alpha\hbar\omega}{\left(\sqrt{\frac{m\omega}{\hbar}}\, x\right)^2} \qquad x \geq 0.$$

Treating

$$V = \frac{\alpha\hbar\omega}{\left(\sqrt{\frac{m\omega}{\hbar}}\, x\right)^2}$$

as a perturbation, study the energy of the ground state of this system.

The ground state energy of the unperturbed Hamiltonian is 3/2 $\hbar\omega$. The first-order energy $E_0^{(1)} \equiv V_{00}$ is

$$\alpha\hbar\omega \int \phi_0^* \frac{1}{\left(\sqrt{\frac{m\omega}{\hbar}}\, x\right)^2} \phi_0 \, dx = \alpha 2\hbar\omega$$

where

$$\phi_0 = \left(\frac{m\omega}{\hbar}\right)^{1/4} \frac{1}{\pi^{1/4}} e^{-u^2/2}\, 2u \quad \text{and} \quad u = \sqrt{\frac{m\omega}{\hbar}}\, x \, .$$

The second order energy $E_0^{(2)} \equiv \sum_{m\neq 0} \frac{V_{0m}V_{m0}}{\varepsilon_0 - \varepsilon_m}$ is

$$\frac{|\langle 0|V|1\rangle|^2}{-2\hbar\omega} + \frac{|\langle 0|V|2\rangle|^2}{-4\hbar\omega} + \frac{|\langle 0|V|3\rangle|^2}{-6\hbar\omega} + \ldots$$

$$= -\frac{4}{3}\alpha^2\hbar\omega \left\{1 + \frac{2}{5} + \frac{8}{35} + \ldots \right\} .$$

This is a slowly converging series.

Here

$$\phi_1 = \left(\frac{m\omega}{\hbar}\right)^{1/4} \frac{2}{\sqrt{6}\pi^{\frac{1}{4}}} \; e^{-u^2/2}\, u(2u^2 - 3),$$

$$\phi_2 = \left(\frac{m\omega}{\hbar}\right)^{1/4} \frac{1}{\sqrt{30}\pi^{\frac{1}{4}}} \; e^{-u^2/2}\, u(4u^4 - 20u^2 + 15)$$

$$\phi_3 = \left(\frac{m\omega}{\hbar}\right)^{1/4} \frac{1}{\sqrt{1260}\pi^{\frac{1}{4}}} \; e^{-u^2/2}\, u(8u^6 - 84u^4 + 210u^2 - 105).$$

CHAPTER 12

Degeneracy

One dimensional systems

The Schrödinger eigenvalue Equation (12.1) for the energy of a system in the simplest possible case

$$\left\{- \frac{\hbar^2}{2M} \frac{d^2}{dx^2} + V(x)\right\}\psi(x) = E\psi(x), \tag{12.1}$$

namely that of a one dimensional system, involves a second-order differential equation. Hence there should in general be two solutions ψ_1, ψ_2 for each energy E. Such two-fold 'degeneracy' as it is called does not usually arise however. It is removed by the boundary conditions.

Consider for instance a particle in the well:

$$V = \infty \quad x < 0, \qquad V = -|V_0| \quad 0 < x < a, \qquad V = 0 \quad x > a,$$

illustrated in Figure 12.1, where $E < 0$.

In the region $0 < x < a$ one has two solutions

$$\psi(x) = A \sin Kx \quad \text{or} \quad B \cos Kx, \qquad K = \sqrt{\frac{2M\{|V_0|-|E|\}}{\hbar^2}},$$

while in the region $a < x < \infty$

$$\psi(x) = Ce^{-kx} \quad \text{or} \quad De^{kx}, \quad k = \sqrt{\frac{2m|E|}{\hbar^2}}.$$

However, the wavefunction in this case satisfies the boundary condition that it vanishes at $x = 0$ where the potential is infinitely repulsive, and at infinity since the particle is localized in the well.

Hence $B \cos Kx$ and De^{kx} must be discarded. One thus has only <u>one</u> acceptable solution which together with its derivative must be made continuous for all $x > 0$, in particular at $x = a$. This in turn imposes a quantization condition on the energy namely

$$\tan Ka = -\frac{K}{k}\ .$$

In two or three dimensional systems the Schrödinger eigenvalue equation may involve degeneracies.

Two dimensional systems

Consider the case where $V = V(\rho)$. The Schrödinger eigenvalue equation is then (see Equation (8.10)).

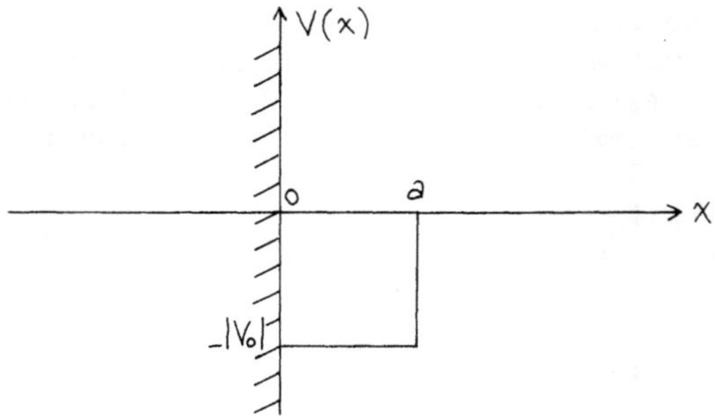

Figure 12.1. Potential producing non-degenerate eigenfunctions.

$$\left\{-\frac{\hbar^2}{2M}\frac{d^2}{d\rho^2} + V(\rho) + \frac{\hbar^2(m^2-\frac{1}{4})}{2M\rho^2}\right\} w_{nm}(\rho) = E_{nm}w_{nm}(\rho), \tag{12.2}$$

and the total wavefunction of the system is:

$$\chi_{n\pm m}(\rho,\ \phi) = \frac{w_{nm}(\rho)}{\rho^{1/2}}\ \frac{e^{\pm im\phi}}{\sqrt{2\pi}}$$

Independent of the details of $w_{nm}(\rho)$ there are here at least two normalized solutions for each energy eigenvalue, unless $m = 0$ namely

$$\frac{w_{nm}(\rho)}{\rho^{1/2}}\ \frac{e^{im\phi}}{\sqrt{2\pi}},\quad \frac{w_{nm}(\rho)}{\rho^{1/2}}\ \frac{e^{-im\phi}}{\sqrt{2\pi}} \tag{12.3}$$

where

$$\int_0^\infty w_{nm}^2(\rho)\ d\rho = 1),$$

because Equation (12.2) depends on m^2 rather than m. One way this m degeneracy can be removed is by inserting a term in the interaction which depends on the variable ϕ for instance the term αL_ϕ which adds the linear term $\alpha\hbar m$ to the operator in square brackets in Equation (12.2) and distinguishes and differentiates between the +m and -m cases. Additional degeneracies may arise depending on the details of the potential $V(\rho)$ provided more than one n, m combination corresponds to a given energy.

EXAMPLE 12.1. Consider the two-dimensional oscillator potential $V = \frac{1}{2}M\omega^2\rho^2$. Solving the problem in cylindrical coordinates yields $E_{nm} = (2n+|m|+1)\hbar\omega$ (see Equation (8.14)). A simple tabulation shows the degeneracy increases as the energy increases, in fact is N, where $N\hbar\omega$ is the energy of the system. These results are tabulated in Table 12.1.

It is of interest to compare these with the degeneracies which result if one solves the same problem in Cartesian coordinates. This particular potential can also be written $\frac{1}{2}M\omega^2(x^2 + y^2)$ and has solutions:

$$\psi_{n_x n_y}(x, y) = \phi_{n_x}(x)\phi_{n_y}(y),$$

$$E_{n_x n_y} = (n_x + n_y + 1)\hbar\omega = E_{n_x} + E_{n_y} = N\hbar\omega,$$

where

$$\phi_n(x) = \frac{1}{2^{n/2}}\frac{1}{(n!)^{1/2}}\left(\frac{1}{b\sqrt{\pi}}\right)^{1/2}\exp\left(-\frac{x^2}{2b^2}\right)H_n\left(\frac{x}{b}\right) \qquad (12.4)$$

$$E_n = (n + \tfrac{1}{2})\,\hbar\omega, \quad n = 0, 1\ldots \quad b = \sqrt{\frac{\hbar}{M\omega}}\,.$$

Table 12.2, which is similar to Table 12.1 shows that the number of degeneracies for a given energy are, as they must be basis independent.

In the case there is no degeneracy (the ground state) the wave functions are identical. Thus if $E = \hbar\omega$

$$\chi_{00}(\rho, \phi) = \frac{w_{00}(\rho)}{\sqrt{2\pi}\,\rho^{1/2}} = \left(\frac{2}{b^2}\right)^{1/2} e^{-\rho^2/2b^2}\frac{1}{\sqrt{2\pi}} =$$

$$= \left(\frac{1}{b\sqrt{\pi}}\right)^{1/2} e^{-x^2/2b^2}\left(\frac{1}{b\sqrt{\pi}}\right)^{1/2} e^{-y^2/2b^2} = \left(\frac{1}{b\sqrt{\pi}}\right) e^{-\rho^2/2b^2}\,.$$

However in the cases there is degeneracy the wavefunctions are not identical. Thus the $E = 2\,\hbar\omega$ states are

$$\chi_{0\pm1}(\rho, \phi) = \frac{w_{01}(\rho)e^{\pm i\phi}}{\rho^{1/2}\,\sqrt{2\pi}} = \left(\frac{1}{\pi}\right)^{1/2}\frac{\rho}{b^2}\,e^{-\rho^2/2b^2}\,e^{\pm i\phi}$$

Table 12.1. Energy levels in Example 12.1, using cyclindrical coordinates

E_{nm}	n	m	Degeneracy	
			Partial	Total
$1\hbar\omega$	0	0	1	1
$2\hbar\omega$	0	±1	2	2
$3\hbar\omega$	1	0	1	3
	0	±2	2	
$4\hbar\omega$	1	±1	2	4
	0	±3	2	

Table 12.2. Energy levels in Example 12.1 using Cartesian coordinates

$E_{n_x n_y}$	N	n_x	n_y	Degeneracy
$1\hbar\omega$	1	0	0	1
$2\hbar\omega$	2	1	0	2
		0	1	
$3\hbar\omega$	3	2	0	3
		1	1	
		0	2	
$4\hbar\omega$	4	3	0	4
		2	1	
		1	2	
		0	3	

in cylindrical coordinates while in Cartesian coordinates:

$$\psi_{01}(x, y) = \left(\frac{1}{b\sqrt{\pi}}\right)^{1/2} \exp\left(-x^2/2b^2\right) \frac{1}{\sqrt{2}}\left(\frac{1}{b\sqrt{\pi}}\right)^{1/2} \exp\left(-y^2/2b^2\right)\frac{2y}{b} =$$

$$= \left(\frac{2}{\pi}\right)^{1/2} \frac{y}{b^2} e^{-\rho^2/2b^2}$$

and

$$\psi_{10}(x, y) = \left(\frac{2}{\pi}\right)^{1/2} \frac{x}{b^2} e^{-\rho^2/2b^2}$$

which are obviously not identical with $\chi_{0\pm1}(\rho, \phi)$.
Despite this one can quickly confirm that

$$\chi_{0\pm1}(\rho, \phi) = \frac{\psi_{10}(x, y) \pm i\psi_{01}(x, y)}{\sqrt{2}} .$$

Generally

$$\chi_{nm}(\rho, \phi) = \sum_{n_x, n_y} a_{n_x n_y} \psi_{n_x n_y}(x, y) \tag{12.5}$$

where the sum is over all degenerate states at the same (in this case $n_x + n_y = 2n+|m|$) energy as $\chi_{nm}(\rho, \phi)$ and reciprocally:

$$\psi_{10}(x, y) = \frac{\chi_{01}(\rho, \phi) + \chi_{0-1}(\rho, \phi)}{\sqrt{2}} ,$$

$$\psi_{01}(x, y) = \frac{\chi_{01}(\rho, \phi) - \chi_{0-1}(\rho, \phi)}{i\sqrt{2}}$$

i.e.

$$\psi_{n_x n_y}(x, y) = \sum_{n, m} b_{nm} \chi_{nm}(\rho, \phi), \tag{12.6}$$

with analogous restrictions on n, m in this sum.

EXAMPLE 12.2. Discuss the degeneracies for a particle in a two-dimensional Coulomb potential (cf. Equation (8.12)).
The energy for this system is

$$E_{nm} = - \frac{Z^2\alpha^2 Mc^2}{2n^2} , \quad n = \frac{1}{2}, \frac{3}{2} \ldots$$

The degeneracy is

$$2n = 1 + \sum_{m=1}^{n-\frac{1}{2}} 2$$

since, aside from m = 0 all other terms are two-fold degenerate, while m varies from 0 to $n - \frac{1}{2}$.

Three-dimensional systems

Consider the case V = V(r). The Schrödinger eigenvalue equation is then (see Equation (8.2))

$$\left\{-\frac{\hbar^2}{2M}\frac{d^2}{dr^2} + \left(V(r) + \frac{\hbar^2 \ell(\ell+1)}{2Mr^2}\right)\right\} u_{n\ell}(r) = E_{n\ell} u_{n\ell}(r)$$

where (12.7)

$$\psi_{n\ell m}(r, \theta, \phi) = \frac{u_{n\ell}(r)}{r} Y_m^\ell (\theta, \phi) .$$

This equation has at least a $(2\ell+1)$ fold degeneracy because Equation (12.7) is independent of the integer m which can take all values from $-\ell$ to ℓ. There are additional degeneracies which depend on the details of V(r), i.e. how many different n, ℓ combinations yield the same energy $E_{n\ell}$.

EXAMPLE 12.3. Discuss the degeneracies of the three-dimensional simple harmonic oscillator.

For the three-dimensional simple harmonic oscillator the energies are given by $E_{n\ell} = (2n + \ell + 3/2)\hbar\omega$ (see Equation (8.13)), in spherical coordinates and $E_{n_x n_y n_z} = (n_x + n_y + n_z + 3/2)\hbar\omega$ in Cartesian coordinates. One can easily verify that if $E = (N+3/2)\hbar\omega$ the degeneracy is

$$(N+1)(N+2)/2 \text{ fold} = \sum_{\substack{\ell=0 \\ \text{even}}}^{N} (2\ell+1) \text{ if N is even or}$$

$$\sum_{\substack{\ell=1 \\ \text{odd}}}^{N} (2\ell+1)$$

if N is odd where $N = n_x + n_y + n_z = 2n + \ell$

EXAMPLE 12.4. Discuss the degeneracies of the three-dimensional Coulomb potential. For the three-dimensional Coulomb problem $E_n = - 1/2\, Mc^2\alpha^2 1/n^2$ (see Equation (8.11)) and the degeneracy is $\Sigma_{\ell=0}^{n-1} (2\ell+1) = n^2$. This considerable degeneracy in the three-dimensional Coulomb system is due to the fact that 'accidentally' the energy in this case does not depend on ℓ. This system has a so called 'accidental degeneracy' in addition to the degeneracy due to the fact that the energy does not depend on m. In fact the Nth energy level of the Coulomb system with a degeneracy N^2 has approximately twice the degeneracy of the three-dimensional simple harmonic oscillator's ($\sim N^2/2$ for large N) Nth energy level. This is exactly true about the degeneracy of the Nth energy level of the two dimensional Coulomb potential = 2N as compared to that of the two-dimensional harmonic oscillator potential = N. The two-dimensional Coulomb potential also has an 'accidental degeneracy' in that the energy is completely independent of m.

In the three dimensional problem one may remove the $(2\ell+1)$ degeneracy of each state for example by adding a term of the form αL_z to the Hamiltonian, and hence a term $\alpha\hbar m$ to the operator in brackets in Equation (12.7).

If one has two or more particle systems the degeneracies are generally quite numerous.

EXAMPLE 12.5. Consider the system

$$H(1, 2) = L^2(1) + L^2(2)$$

$$E_{\ell_1\ell_2} = \hbar^2[\ell_1(\ell_1+1) + \ell_2(\ell_2+1)]$$

$$\psi^{\ell_1\ell_2}_{m_1m_2}(1, 2) = \frac{1}{\sqrt{2}}\left(Y^{\ell_1}_{m_1}(1)\, Y^{\ell_2}_{m_2}(2) \pm Y^{\ell_1}_{m_1}(2)\, Y^{\ell_2}_{m_2}(1)\right)$$

(where the $\frac{1}{\sqrt{2}}$ is inserted in $\psi^{\ell_1\ell_2}_{m_1m_2}(1, 2)$ for normalization purposes).

For a given ℓ_1, ℓ_2 there are here generally $(2\ell_1+1)(2\ell_2+1)$ symmetric wavefunctions and the same number of antisymmetric wavefunctions where for symmetric functions $\psi(1, 2) = \psi(2, 1)$ while for antisymmetric wavefunctions $\psi(1, 2) = -\psi(2, 1)$. In the special case $\ell_1 = \ell_2$ there are only $(2\ell+1)^2$ wavefunctions at the energy $2\hbar^2\ell(\ell+1)$ of which $2\ell(2\ell+1)$ are symmetric and $(2\ell+1)$ antisymmetric. These results can be easily illustrated by listing the wavefunctions for $\ell_1 = 1$, $\ell_2 = 1$; $\ell_1 = 1$, $\ell_2 = 2$.

(a) $\ell_1 = 1$, $\ell_2 = 1$ $E = 4\hbar^2$,

$$\psi(1, 2) = Y^1_1(1)Y^1_1(2);\ \frac{1}{\sqrt{2}}[Y^1_1(1)Y^1_0(2) \pm Y^1_1(2)Y^1_0(1)];$$

$$\frac{1}{\sqrt{6}}[2Y^1_0(1)Y^1_0(2) + Y^1_1(1)Y^1_{-1}(2) + Y^1_1(2)Y^1_{-1}(1)];$$

$$\frac{1}{\sqrt{3}}[Y^1_0(1)Y^1_0(2) - Y^1_1(1)Y^1_{-1}(2) - Y^1_1(2)Y^1_{-1}(1)];$$

$$\frac{1}{\sqrt{2}}[Y^1_1(1)Y^1_{-1}(2) - Y^1_1(2)Y^1_{-1}(1)];$$

$$\frac{1}{\sqrt{2}}[Y^1_{-1}(1)Y^1_0(2) \pm Y^1_{-1}(2)Y^1_0(1)];\ Y^1_{-1}(1)Y^1_{-1}(2),$$

i.e. a ninefold degeneracy. (This particular choice corresponds to the set of eigenfunctions ψ^{ℓ}_{m} of the total angular momentum operator $\vec{L} = \vec{L}(1) + \vec{L}(2)$ where ℓ = 2, 1, and 0)

(b) $\ell_1 = 1,\ \ell_2 = 2,\ E = 8\hbar^2$

$$\psi(1, 2) = \frac{1}{\sqrt{2}}\left\{Y^1_{\pm1}(1)Y^2_{\pm2}(2) \pm Y^1_{\pm1}(2)Y^2_{\pm2}(1)\right\};\ \frac{1}{\sqrt{2}}\left\{Y^1_0(1)Y^2_{\pm2}(2) \pm Y^1_0(2)Y^2_{\pm2}(1)\right\};$$

$$\frac{1}{\sqrt{2}}\left\{Y^1_{\mp1}(1)Y^2_{\pm2}(2) \pm Y^1_{\mp1}(2)Y^2_{\pm2}(1)\right\};\ \frac{1}{\sqrt{2}}\left\{Y^1_{\pm1}(1)Y^2_{\pm1}(2) \pm Y^1_{\pm1}(2)Y^2_{\pm1}(1)\right\};$$

$$\frac{1}{\sqrt{2}}\left\{Y^1_{\mp1}(1)Y^2_{\pm1}(2) \pm Y^1_{\mp1}(2)Y^2_{\pm1}(1)\right\};\ \frac{1}{\sqrt{2}}\left\{Y^1_{\pm1}(1)Y^2_0(2) \pm Y^1_{\pm1}(2)Y^2_0(1)\right\};$$

$$\frac{1}{\sqrt{2}}\left\{Y^1_0(1)Y^2_{\pm1}(2) \pm Y^1_0(2)Y^2_{\pm1}(1)\right\};\ \frac{1}{\sqrt{2}}\left\{Y^1_0(1)Y^2_0(2) \pm Y^1_0(2)Y^2_0(1)\right\},$$

i.e. a 30 fold degeneracy,

The fact one can write these eigenfunctions as either symmetric or antisymmetric combinations is due to the fact that if P_{12} is an exchange operator where $P_{12}\eta(1, 2) = \eta(2, 1)$ the Hamiltonian in this problem commutes with P_{12} i.e. $[P_{12}, H_{12}] = 0$. But the eigenvalues λ of P_{12} are ±1 since if $P_{12}\eta(1, 2) = \lambda\eta(1, 2) = \eta(2, 1)$, then $P^2_{12}\eta(1, 2) = \lambda P_{12}\eta(1, 2) = \lambda^2\eta(1, 2) = P_{12}\eta(2, 1) = \eta(1, 2)$ i.e. $\lambda^2 = 1$, $\lambda = \pm1$. Since P_{12} and H(1, 2) commute this means the eigenfunctions of H can also be written as eigenfunctions of P_{12} i.e. combinations which have eigenvalues +1 or -1 under particle interchange.

One can remove the degeneracy of this system for instance by adding the term $\alpha_1 L_z(1) + \alpha_2 L_z(2)$ to the Hamiltonian. With this term the Hamiltonian no longer commutes with the exchange operator (unless $\alpha_1 = \alpha_2$) and in addition different projections have different energies so there is no degeneracy. If $\alpha_1 = \alpha_2$ there is still 'exchange' degeneracy, though states with different projections are no longer degenerate. In the case above where $\ell_1 = \ell_2 = 1$ one then has one state with energy $4\hbar^2 + 2\alpha\hbar$ two states with energy $4\hbar^2 + \alpha\hbar$, three states with energy $4\hbar^2$, two states with energy $4\hbar^2 - \alpha\hbar$ and one state with energy $4\hbar^2 - 2\alpha\hbar$.

EXAMPLE 12.6. Consider $H(1, 2) = T_1 + T_2 + \frac{1}{2}M\omega^2 r_1^2 + \frac{1}{2}M\omega^2 r_2^2$.
Discuss the degeneracies of this two particle system,

$$E = (2n_1 + \ell_1 + \frac{3}{2} + 2n_2 + \ell_2 + \frac{3}{2})\hbar\omega .$$

The degeneracies of the lowest three energy levels are shown in Table 12.3. One notes the degeneracies are quite numerous and that symmetric and anti-symmetric combinations of the resulting wavefunctions can easily be formed.
Thus for $E = 4\hbar\omega$:

$$\psi_\pm(1, 2) = \frac{1}{\sqrt{2}} \left\{R_{01}(1)R_{00}(2)Y_{1m}(1)Y_{00}(2) \pm R_{00}(1)R_{01}(2)Y_{00}(1)Y_{1m}(2)\right\}$$

etc,

where ± stand for symmetric and antisymmetric wavefunctions respectively and the $1/\sqrt{2}$ is inserted to ensure wavefunction normalization.

EXAMPLE 12.7. Consider a particle subject to the Hamiltonian

$$H = \ell n(1 + \frac{\alpha}{\hbar^2} (L_x^2 + L_y^2)) .$$

Discuss the energy levels of this system and their degeneracies.

$$L_x^2 + L_y^2 = L^2 - L_z^2$$

is a useful identity here.
Writing: $H\psi = \ell n(1 + \alpha/\hbar^2 (L^2 - L_z^2))\psi = E\psi$ one obtains

$$E = \ell n(1+\alpha(\ell(\ell+1)-m^2)), \quad \psi = Y_m^\ell(\theta, \phi), \quad -\ell \le m \le \ell$$

$$\ell = 0, 1, 2...$$

$$m = \text{integer}$$

The first few levels and their degeneracies are illustrated in Table 12.4.

One notes that one can write for E, $E = \ell n(1+n\alpha)$, $n = 0, 1, 2,...$ with different degeneracies for different values of n.
Finally for small α $E \sim \alpha(\ell(\ell+1) - m^2)$.
Adding a term proportional to L_z will remove some or all the degeneracy depending on the coefficient in front of this additional term.

When one has an energy state n which is say j fold degenerate $(n_1, n_2, ..., n_j)$ it may not be possible to use the perturbation expansion (11.14) directly. In particular if the perturbation V connects any of the j degenerate states say $n = n_1$ and $m = n_i$ $1 < i \le j$ i.e.

$V_{n_1 n_i} = \langle n_1|V|n_i\rangle \neq 0$, there will be terms in this expansion with a zero energy denominator $\varepsilon_{n_1} - \varepsilon_{n_i}$ and non-zero numerator $V_{n_1 n_i}$.

Table 12.3. Degeneracies of Hamiltonian in Example 12.6.

Energy	Degeneracy	n_1	n_2	ℓ_1	ℓ_2	$\psi(1, 2)$	m_1	m_2
$3\hbar\omega$	None	0	0	0	0	$R_{00}(1)R_{00}(2)Y_{00}(1)Y_{00}(2)$	0	0
$4\hbar\omega$	3+3 =	0	0	1	0	$R_{01}(1)R_{00}(2)Y_{1m_1}(1)Y_{00}(2)$	0,±1	0
	six fold	0	0	0	1	$R_{00}(1)R_{01}(2)Y_{00}(1)Y_{1m_2}(2)$	0	0,±1
	9+5+5+1+1=	0	0	1	1	$R_{01}(1)R_{01}(2)Y_{1m_1}(1)Y_{1m_2}(2)$	0,±1	0,±1
		0	0	0	2	$R_{00}(1)R_{02}(2)Y_{00}(1)Y_{2m_2}(2)$	0	0,±1,±2
$5\hbar\omega$	twenty-one fold	0	0	2	0	$R_{02}(1)R_{00}(2)Y_{2m_1}(1)Y_{00}(2)$	0,±1,±2	0
		0	1	0	0	$R_{00}(1)R_{10}(2)Y_{00}(1)Y_{00}(2)$	0	0
		1	0	0	0	$R_{10}(1)R_{00}(2)Y_{00}(1)Y_{00}(2)$	0	0

Table 12.4. Degeneracies of Hamiltonian in Example 12.7.

E	ℓ	m	Degeneracy
0	0	0	1
$\ell n(1+\alpha)$	1	±1	2
$\ell n(1+2\alpha)$	1	0	1 } 3
	2	±2	2
$\ell n(1+3\alpha)$	3	±3	2
$\ell n(1+4\alpha)$	4	±4	2
$\ell n(1+5\alpha)$	2	±1	2 } 4
	5	±5	2
$\ell n(1+6\alpha)$	2	0	1 } 3
	6	±6	2
$\ell n(1+7\alpha)$	7	±7	2

In this case expansion (11.14) will consequently be divergent. The way out of this difficulty is to use the fact (see Equation (12.5) and (12.6)) that one can take different combinations of the j degenerate functions, in particular a combination for which all $V_{n_i n_{i'}}$, $i \neq i'$, where $i, i' \leq j$ are zero. To do this one diagonalizes V in each subapace of degenerate states. The resulting eigenfunctions will have only diagonal matrix elements of V. If one therefore uses this basis in expansion (11.14) it will not diverge since terms with zero denominator also have zero numerator. Moreover the eigenvalues of these diagonalizations are just the first-order energies of each state n since $E = \varepsilon_n + V_{nn}$ to first order and only the diagonal elements contribute to V_{nn} in this new basis.

In the simple case of two degenerate states n_1 and n_2 connected by V, the normalized combination one must choose (which uncouples if $\phi = 0$) is

$$|n_1'\rangle = |n_1\rangle \cos\phi + |n_2\rangle \sin\phi$$

$$|n_2'\rangle = -|n_1\rangle \sin\phi + |n_2\rangle \cos\phi , \qquad (12.8)$$

with a ϕ which makes $\langle n_1'|V|n_2'\rangle = 0$ namely:

$$\tan 2\phi = \frac{2V_{12}}{V_{11}-V_{22}} \quad \text{where } V_{12} = \langle n_1|V|n_2\rangle \text{ etc,} \qquad (12.9)$$

(provided $V_{12} = V_{21}$).

If one has three degenerate states n_1, n_2, and n_3 connected by V, a normalized combination (which uncouples when θ, $\phi = 0$) which one may choose is:

$$|n_1'\rangle = \cos\theta \cos\phi|n_1\rangle + \cos\theta \sin\phi|n_2\rangle + \sin\theta|n_3\rangle$$

$$|n_2'\rangle = -\sin\phi|n_1\rangle + \cos\phi|n_2\rangle \qquad (12.10)$$

$$|n_3'\rangle = \sin\theta\cos\phi|n_1\rangle + \sin\theta\sin\phi\,|n_2\rangle - \cos\theta|n_3\rangle$$

where $\langle n_1'|V|n_2'\rangle$, $\langle n_1'|V|n_3'\rangle$ and $\langle n_2'|V|n_3'\rangle$ are zero, (Provided $V_{12} = V_{21}$, $V_{13} = V_{31}$, $V_{23} = V_{32}$)

i.e.

$$\tan 2\phi = \frac{2V_{12}}{V_{11}-V_{22}}$$

$$\tan 2\theta = \frac{2(V_{13}\cos\phi + V_{23}\sin\phi)}{V_{11}\cos^2\phi + V_{22}\sin^2\phi + V_{12}\sin 2\phi - V_{33}} . \qquad (12.11)$$

EXAMPLE 12.8. Consider a two dimensional oscillator

$$H_0 = \tfrac{1}{2}m\omega^2 (x^2+y^2) + \frac{p_x^2}{2m} + \frac{p_y^2}{2m} , \quad V = V_0\delta(x)\delta(y),$$

where one wishes to find the energy of the second excited state ($E = (N+1)\hbar\omega$, $N = 2$ if $V_0 = 0$).

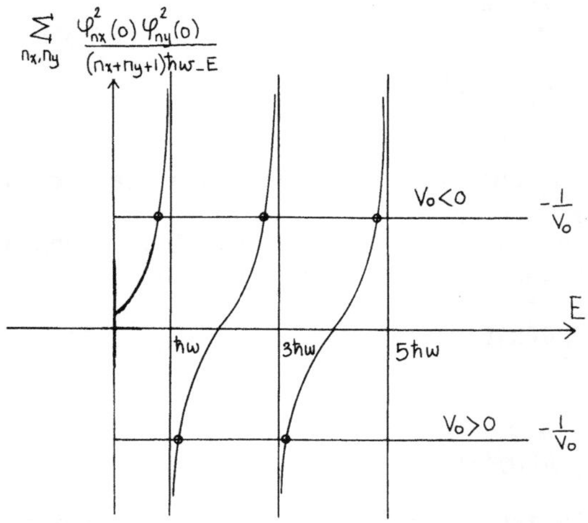

Figure 12.2. Graphical solution for energy levels in Example 12.8.

Clearly the Cartesian basis is convenient here where the basis states for N = 2 are $|2_1) = \phi_2(x)\phi_0(y)$, $|2_2) = \phi_1(x)\phi_1(y)$, $|2_3) = \phi_0(x)\phi_2(y)$, (see expression (12.4)).
Using Equation (12.11), $\tan 2\phi = 2V_{12}/(V_{11}-V_{22}) = 0$ i.e. $\phi = 0$ since $\phi_1(0) = 0$,

$$\tan 2\theta = \frac{2V_{13}}{V_{11}-V_{33}} = \frac{2\phi_2^2(0)\ \phi_0^2(0)}{\phi_2^2(0)\ \phi_0^2(0) - \phi_0^2(0)\ \phi_2^2(0)} = \infty$$

i.e. $2\theta = \pi/2$, $\theta = \pi/4$.

$$|2_1') = \frac{1}{\sqrt{2}}\left\{\phi_2(x)\phi_0(y) + \phi_0(x)\phi_2(y)\right\}$$

$$|2_2') = \phi_1(x)\ \phi_1(y) \qquad (12.12)$$

$$|2_3') = \frac{1}{\sqrt{2}}\left\{\phi_2(x)\ \phi_0(y) - \phi_0(x)\ \phi_2(y)\right\}$$

and the energy to first order is

$$E = \begin{cases} 3\hbar\omega + \langle 2_1'|V|2_1'\rangle = 3\hbar\omega + 2V_0\phi_0^2(0)\ \phi_2^2(0) = 3\hbar\omega + \dfrac{V_0 m\omega}{\pi\hbar} \\ 3\hbar\omega + \langle 2_2'|V|2_2'\rangle = 3\hbar\omega \\ 3\hbar\omega + \langle 2_3'|V|2_3'\rangle = 3\hbar\omega. \end{cases} \tag{12.13}$$

Two of the three basis states thus remain degenerate states to first order in perturbation theory.

This problem can also be solved exactly by graphical techniques. Thus the Schrödinger Equation for this system can be written

$$(H_0 - E)\psi(x,\ y) = -V_0\delta(x)\delta(y)\psi(x,\ y).$$

Hence

$$\psi(x,\ y) = -V_0\,\frac{1}{H_0 - E}\,\delta(x)\delta(y)\psi(x,\ y).$$

Using closure one may insert a complete set of states which are eigenfunctions of H_0 (namely $\sum_{n_x, n_y} \phi_{n_x}(x)\ \phi_{n_y}(y)$) between the operator $1/(H_0-E)$ and $\delta(x)\delta(y)$ obtaining:

$$\psi(x,\ y) = -V_0\,\frac{1}{H_0 - E}\sum_{n_x, n_y} |\phi_{n_x}(x)\phi_{n_y}(y)\rangle\langle\phi_{n_x}(x)\phi_{n_y}(y)\delta(x)\delta(y)\psi(x,\ y)\rangle$$

$$= -V_0\sum_{n_x, n_y}\frac{1}{(n_x + n_y + 1)\hbar\omega - E}\,|\phi_{n_x}(x)\phi_{n_y}(y)\rangle\ \phi_{n_x}(0)\phi_{n_y}(0)\psi(0,\ 0).$$

Inserting the values x, y = 0 into this expression yields;

$$-\frac{1}{V_0} = \sum_{n_x, n_y}\frac{1}{(n_x + n_y + 1)\hbar\omega - E}\,\phi_{n_x}^2(0)\phi_{n_y}^2(0),$$

i.e.,

$$-\frac{1}{V_0} = \frac{\phi_0^4(0)}{\hbar\omega - E} + \frac{2\phi_0^2(0)\phi_2^2(0)}{3\hbar\omega - E} + \frac{(2\phi_0^2(0)\phi_4^2(0) + \phi_2^4(0))}{5\hbar\omega - E} + \ldots$$

This expression is illustrated graphically in Figure 12.2. Consider the energy level close to $3\hbar\omega$, $E \simeq 3\ \hbar\omega + \varepsilon$. Approximately for this level

$$-\frac{1}{V_0} = \frac{2\phi_0^2(0)\ \phi_2^2(0)}{-\ \varepsilon} \quad \text{i.e.} \quad \varepsilon = 2V_0\phi_0^2(0)\ \phi_2^2(0),$$

and $E = 3\hbar\omega + 2V_0\ \phi_0^2(0)\phi_2^2(0)$, in agreement with the degenerate perturbation theory result of Equation (12.13).

EXAMPLE 12.9. Consider the two dimensional oscillator

$$H = \frac{1}{2} m\omega^2(x^2+y^2) + \frac{p_x^2}{2m} + \frac{p_y^2}{2m}$$

$$V = \lambda m\omega^2 xy$$

where one wishes to find the energy of the first excited state (If $\lambda = 0$, $E = (N+1)\hbar\omega$ where $N = 1$).

Clearly the Cartesian basis is convenient here where the basis states are

$$|1_1) = \phi_1(x)\ \phi_0(y) \text{ (see expression (12.4)).}$$

$$|1_2) = \phi_0(x)\phi_1(y)$$

Using the results given by Equation (12.9)

$$\tan 2\phi = \frac{2V_{12}}{V_{11} - V_{22}} = \infty \quad \text{since } V_{11} = V_{22} = 0 \text{ whereas}$$

$$V_{12} \neq 0 = \lambda m\omega^2 < \phi_1(x)\ \phi_0(y)|xy|\phi_0(x)\phi_1(y)> =$$

$$= \lambda m\omega^2 <\phi_1(x)|x|\phi_0(x)>^2$$

i.e. $\phi = \pi/4$

Thus

$$|1_1'> = \frac{1}{\sqrt{2}} |(\phi_1(x)\phi_0(y) + \phi_0(x)\phi_1(y))> \tag{12.14}$$

$$|1_2'> = -\frac{1}{\sqrt{2}} |(\phi_1(x)\phi_0(y) - \phi_0(x)\ \phi_1(y))>$$

and the energy to first order is

$$2\hbar\omega + <1_1'|V|1_1'> = 2\hbar\omega+\lambda m\omega^2<\phi_1(x)|x|\phi_0(x)>^2 = 2\hbar\omega+\lambda m\omega^2<\phi_0(x)|x^2|\phi_0(x)>$$

$$= 2\hbar\omega+2\lambda<\phi_0(x)|\tfrac{1}{2}m\omega^2x^2|\phi_0(x)> = 2\hbar\omega+\lambda\hbar\omega/2. \tag{12.15}$$

E =

$$2h\omega + <1_2'|V|1_2'> = 2\hbar\omega-\lambda m\omega^2<\phi_1(x)|x|\phi_0(x)>^2$$

$$= 2\hbar\omega - \lambda\hbar\omega/2.$$

This problem can also be solved exactly since

$$H+V = \frac{1}{2}m(\dot{x}^2+\dot{y}^2)+\frac{1}{2}m\omega^2(x^2+2\lambda xy+y^2) = \frac{1}{2}m(\dot{X}^2+\dot{Y}^2)+\frac{1}{2}m\omega^2(1-\lambda)X^2+\frac{1}{2}m\omega^2(1+\lambda)Y^2$$

where

$$X = \frac{1}{\sqrt{2}}(x-y), \quad Y = \frac{1}{\sqrt{2}}(x+y) \quad \text{and} \quad \dot{\xi} = \frac{d\xi}{dt}.$$

Written in terms of X, Y the Hamiltonian decouples and is seen to have eigenvalues:

$$E_{N_X N_Y} = (N_X+\tfrac{1}{2})\hbar\omega\sqrt{1-\lambda} + (N_Y+\tfrac{1}{2})\hbar\omega\sqrt{1+\lambda}. \qquad (12.16)$$

To order λ one thus has

$$E_{01} = 2\hbar\omega + \frac{\lambda\hbar\omega}{2}$$

$$E_{10} = 2\hbar\omega - \frac{\lambda\hbar\omega}{2},$$

etc. in agreement with the degenerate perturbation theory results of Equation (12.15) above.

Likewise the exact wavefunctions corresponding to these energies are:

$$\psi_1^1 = \phi_1(X, \omega=\omega_1)\phi_0(Y, \omega=\omega_2) \quad \text{and} \quad \psi_1^2 = \phi_0(X, \omega=\omega_1)\phi_1(Y, \omega=\omega_2), \qquad (12.17)$$

where

$$\omega_1 = \omega\sqrt{1-\lambda}, \quad \omega_2 = \sqrt{1+\lambda}$$

i.e.

$$\psi_1^1 = \frac{1}{\sqrt{2}}\left(\frac{m\omega}{\hbar\pi}\right)^{1/2}(1-\lambda^2)^{1/8}\, e^{-\frac{m\omega\sqrt{1-\lambda}\, X^2}{2\hbar}}\; 2X\left(\frac{m\omega}{\hbar}\right)^{1/2}(1-\lambda)^{1/4}\, e^{-\frac{m\omega\sqrt{1+\lambda}\, Y^2}{2\hbar}}$$

$$\simeq (\phi_1(x)\phi_0(y) - \phi_0(x)\phi_1(y))/\sqrt{2} + O(\lambda) \quad \text{and}$$

$$\psi_1^2 = \frac{1}{\sqrt{2}}\left(\frac{m\omega}{\hbar\pi}\right)^{1/2}(1-\lambda^2)^{1/8}\, e^{-\frac{m\omega\sqrt{1-\lambda}\, X^2}{2\hbar}}\; 2Y\left(\frac{m\omega}{\hbar}\right)^{1/2}(1+\lambda)^{1/4}\, e^{-\frac{m\omega\sqrt{1+\lambda}\, Y^2}{2\hbar}}$$

$$\simeq (\phi_1(x)\phi_0(y) + \phi_0(x)\phi_1(y))/\sqrt{2} + O(\lambda)$$

consistent with Equation (12.14) to within an unimportant overall phase.

EXAMPLE 12.10. Consider the Hamiltonian $H = -\frac{\hbar^2}{2I}\frac{d^2}{d\phi^2} + V_0\delta(\phi-\phi_0)$

(Here $\int_0^{2\pi} \delta(\phi-\phi_0)\, d\phi = 1,\quad \int_0^{2\pi} f(\phi)\delta(\phi-\phi_0)\, d\phi = f(\phi_0)$).

Treating $-\hbar^2/2I\ d^2/d\phi^2$ as H_0, there are two possible pairs of degenerate eigenfunctions: either $\psi = \sqrt{\frac{1}{\pi}}\cos n\phi$ and $\sqrt{1/\pi}\sin n\phi$ or $\psi = 1/\sqrt{2\pi}\ e^{in\phi}$ and $1/\sqrt{2\pi}\ e^{-in\phi}$, each pair having eigenvalues $\varepsilon_n = (\hbar^2 n^2)/2I$, where n must be an integer so that $\psi(\phi+2\pi)=\psi(\phi)$ i.e. is single valued. Additionally $n \neq 0$ otherwise $\varepsilon_0 = 0$ which violates the uncertainty principle (also for n=0 the wavefunctions are not well behaved being zero or constants).

<u>Choice 1</u>: $n_1 = \sqrt{1/\pi}\cos n\phi$, $n_2 = \sqrt{1/\pi}\sin n\phi$. For this choice,

$$V_{11} = \frac{V_0}{\pi}\cos^2 n\phi_0;\quad V_{12} = \frac{V_0}{2\pi}\sin 2n\phi_0;\quad V_{22} = \frac{V_0}{\pi}\sin^2 n\phi_0,\ \text{and}$$

$$\tan 2\phi = \frac{\frac{2V_0}{2\pi}\sin 2n\phi_0}{\frac{V_0}{\pi}\cos^2 n\phi_0 - \frac{V_0}{\pi}\sin^2 n\phi_0} = \tan 2n\phi_0 \quad \text{i.e. } \phi = n\phi_0.$$

Thus

$$|n_1'\rangle = \frac{1}{\sqrt{\pi}}\left(\cos n\phi\cos n\phi_0 + \sin n\phi\sin n\phi_0\right) = \frac{1}{\sqrt{\pi}}\cos n(\phi-\phi_0)$$

$$|n_2'\rangle = \frac{1}{\sqrt{\pi}}\left(-\cos n\phi\sin n\phi_0 + \sin n\phi\cos n\phi_0\right) = \frac{1}{\sqrt{\pi}}\sin n(\phi-\phi_0)$$

and

$$E_n = \begin{cases} \varepsilon_n + \frac{V_0}{\pi}\int_0^{2\pi}\cos^2 n(\phi-\phi_0)\delta(\phi-\phi_0)\, d\phi = \varepsilon_n + \frac{V_0}{\pi}, \text{ or} \\ \varepsilon_n + \frac{V_0}{\pi}\int_0^{2\pi}\sin^2 n(\phi-\phi_0)\delta(\phi-\phi_0)\, d\phi = \varepsilon_n, \end{cases}$$

a result independent of ϕ_0 i.e. of the choice of original axis orientation.

<u>Choice 2</u>:

$$n_1 = \frac{1}{\sqrt{\pi}} e^{in\phi}, \quad n_2 = \frac{1}{\sqrt{2\pi}} e^{-in\phi}.$$

For this choice,

$$V_{11} = \frac{V_0}{2\pi}; \quad V_{22} = \frac{V_0}{2\pi}; \quad V_{12} = \frac{V_0}{2\pi} e^{-i2n\phi_0}; \quad \text{i.e.} \quad V_{21}^* = V_{12}.$$

Hence we cannot use expression (12.9) which assumes $V_{12} = V_{21}$.

Instead one may have recourse directly to the expression (cf. Equation (12.8)):

$$E_n = \varepsilon_n + \langle n_1^1|V|n_1^1\rangle = \varepsilon_n + V_{11}\cos^2\phi + V_{22}\sin^2\phi + (V_{12}+V_{21})\frac{\sin 2\phi}{2}$$

$$= \varepsilon_n + \frac{V_{11}+V_{22}}{2} + \frac{V_{11}-V_{22}}{2}\cos 2\phi + \frac{V_{12}+V_{21}}{2}\sin 2\phi.$$

Additionally substitution into $\langle n_1'|V|n_2'\rangle = 0$ yields: $(V_{11}-V_{22})\sin 2\phi = (V_{12}-V_{21}) + (V_{12}+V_{21})\cos 2\phi$, from which one obtains

$$\tan 2\phi = [(V_{11}-V_{22})(V_{12}+V_{21}) \pm (V_{12}-V_{21})\sqrt{(V_{11}-V_{22})^2+4V_{12}V_{21}}]/$$

$$[(V_{11}-V_{22})^2 - (V_{12}-V_{21})^2] = \frac{\sin 2\phi}{\cos 2\phi} =$$

$$\frac{[(V_{11}-V_{22})(V_{12}-V_{21})\pm(V_{12}+V_{21})\sqrt{(V_{11}-V_{22})^2+4V_{12}V_{21}}]/[(V_{11}-V_{22})^2+(V_{12}+V_{21})^2]}{[-(V_{12}^2-V_{21}^2)\pm(V_{11}-V_{22})\sqrt{(V_{11}-V_{22})^2-4V_{12}V_{21}}]/[(V_{11}-V_{22})^2+(V_{12}+V_{21})^2]}$$

Hence:

$$E_n = \varepsilon_n + \frac{[(V_{11}+V_{22}) \pm \sqrt{(V_{11}+V_{22})^2 - 4(V_{11}V_{22}-V_{12}V_{21})}]}{2}. \qquad (12.18)$$

Expression (12.18) is quite general. In fact it involves the solutions of the eigenvalue equations which result from diagonalizing the Hamiltonian H in the space of two degenerate states n_1, n_2 i.e.

$$\begin{vmatrix} H_{11}-E_n & V_{12} \\ V_{21} & H_{22}-E_n \end{vmatrix} = 0.$$

Substituting in expression (12.18) the values of V_{11}, V_{22} and V_{12} for choice 2 one obtains:

$$E_n = \varepsilon_n + \frac{V_0}{\pi}\ , \quad E_n = \varepsilon_n$$

which as expected are identical with the energies resulting from choice 1 for the degenerate basis.

EXAMPLE 12.11. For the Hamiltonian in Example 12.10 calculate the second-order energy contribution.

Using the choice 1 for the degenerate states

$$\langle m|V_0\delta(\phi-\phi_0)|n_1'\rangle = \frac{V_0}{\pi}\cos m\,\phi_0,\quad \frac{V_0}{\pi}\sin m\,\phi_0,$$

depending on whether m is $1/\sqrt{\pi}\cos m\phi$ or $1\sqrt{\pi}\sin m\phi$. Also

$$\langle m|V_0\delta(\phi-\phi_0)|n_2'\rangle = 0.$$

Thus

$$E_m^{(2)} = \sum_{\substack{m\neq n\\ \neq 0}} \frac{\langle n_1'|V_0\delta(\phi-\phi_0)|m\rangle\ \langle m|V_0\delta(\phi-\phi_0)|n_1'\rangle}{\varepsilon_n-\varepsilon_m} =$$

$$\sum_{\substack{m\neq n\\ \neq 0}} \left\{ \frac{\frac{V_0^2}{\pi^2}\cos^2 m\phi_0}{\varepsilon_n-\varepsilon_m} + \frac{\frac{V_0^2}{\pi^2}\sin^2 m\phi_0}{\varepsilon_n-\varepsilon_m} \right\} = \frac{V_0^2}{\pi^2} \sum_{\substack{m\neq n\\ \neq 0}} \frac{1}{\varepsilon_n-\varepsilon_m}$$

a result, like in first order independent of ϕ_0 i.e. of the choice of axis orientation. Thus

$$E_m^{(2)} = \frac{2IV_0^2}{\hbar^2\pi^2}\sum_{\substack{m\neq n\\ \neq 0}}\frac{1}{n^2-m^2} = \frac{2IV_0^2}{2n\hbar^2\pi^2}\sum_{\substack{m\neq n\\ \neq 0}}\left\{\frac{1}{m+n} - \frac{1}{m-n}\right\}$$

which can easily be summed for a particular n. Thus if n = 1,

$$E_m^{(2)} = \frac{IV_0^2}{\hbar^2\pi^2}\left\{\frac{1}{3}+\frac{1}{4}+\frac{1}{5}+\ \ldots\ - \left(1+\frac{1}{2}+\frac{1}{3}+\frac{1}{4}+\ \ldots\right)\right\} = -\frac{3IV_0^2}{2\hbar^2\pi^2} \quad \text{etc.}$$

EXAMPLE 12.12. The $2s_{\frac{1}{2}}$ and $2p_{\frac{1}{2}}$ states of the hydrogen atom are separated by an energy ΔE known as the Lamb shift (where $\frac{\Delta E}{\hbar} = 10^9$ Hz). Consider these two states in the presence of an electric dipole field $V = \frac{\lambda mc}{\hbar}z$.

Neglecting the other energy levels of hydrogen calculate exactly the two new eigenstates of the system and the corresponding eigenvalues. Compare

these results with the eigenvalues and energies obtained using non-degenerate and degenerate perturbation theory. (N.B. Consider carefully the parity of the integrals involved.)
Defining:

$$\phi_1 \equiv 2s_{\frac{1}{2}} \equiv |1), \quad \varepsilon_1 \equiv \varepsilon_{2s_{\frac{1}{2}}},$$

$$\phi_2 \equiv 2p_{\frac{1}{2}} \equiv |2), \quad \varepsilon_2 \equiv \varepsilon_{2p_{\frac{1}{2}}},$$

$$V = \frac{mc}{\hbar} z,$$

one obtains

$$\begin{bmatrix} \varepsilon_1 + \lambda(1|V|1) - E & \lambda(1|V|2) \\ \lambda(2|V|1) & \varepsilon_2 + \lambda(2|V|2) - E \end{bmatrix} \begin{bmatrix} a_0 \\ a_1 \end{bmatrix} = 0,$$

in agreement with Equation (11.7).
But $(1|V|1) = (2|V|2) = 0$ from parity considerations.
Therefore

$$E_{1_2} = \frac{\varepsilon_1+\varepsilon_2}{2} \pm \sqrt{\frac{(\varepsilon_1-\varepsilon_2)^2}{4} + \lambda^2(1|V|2)^2}.$$

Also

$$\frac{a_1}{a_0} = \frac{E_1-\varepsilon_1}{\lambda(1|V|2)} = \frac{\left(-\frac{\varepsilon_1-\varepsilon_2}{2} + \sqrt{\frac{(\varepsilon_1-\varepsilon_2)^2}{4} + \lambda^2(1|V|2)^2}\right)}{\lambda(1|V|2)}$$

and

$$\psi_1 = a_0\left(\phi_0 + \frac{a_1}{a_0}\phi_1\right) = a_0\left(\phi_0 + \frac{\left(-\frac{\varepsilon_1-\varepsilon_2}{2} + \sqrt{\frac{(\varepsilon_1-\varepsilon_2)^2}{4} + \lambda^2(1|V|2)^2}\right)}{\lambda(1|V|2)}\phi_1\right) \quad \text{etc.}$$

If

$$\lambda(1|V|2) \ll (\varepsilon_1-\varepsilon_2)/2,$$

$$E_1 \approx \varepsilon_1 + \lambda^2 \frac{(1|V|2)^2}{(\varepsilon_1-\varepsilon_2)} + \ldots$$

and

$$E_2 \approx \varepsilon_2 - \lambda^2 \frac{(1|V|2)^2}{(\varepsilon_1-\varepsilon_2)} + \ldots$$

Also

$$a_1 \approx \frac{\lambda(1|V|2)}{\varepsilon_1-\varepsilon_2}\, a_0$$

$$\psi_1 = a_0 \left\{\phi_0 + \frac{\lambda(1|V|2)}{\varepsilon_1-\varepsilon_2}\ \phi_1\right\}$$

etc., in agreement with non-degenerate perturbation theory for E and ψ.

If $\lambda(1|V|2) >> (\varepsilon_1-\varepsilon_2)/2$, $\varepsilon_1 \simeq \varepsilon_2$ i.e. the Lamb shift is much smaller than the effect of the electric dipole field,

$$E_1 = \frac{\varepsilon_1+\varepsilon_2}{2} + \lambda(1|V|2)\sqrt{1 + \frac{(\varepsilon_1-\varepsilon_2)^2}{4\lambda^2(1|V|2)^2}} \simeq \varepsilon_1 + \lambda(1|V|2)$$

and $a_1/a_0 \approx 1$,

in agreement with the degenerate result (see Equation (12.9)) for the case $\tan 2\phi = \infty$
i.e.

$$\phi = \frac{\pi}{4} \quad \text{in which case } |1') = \frac{1}{\sqrt{2}}(|1) + |2))$$

and

$$E_1 = (1'|H|1') = \varepsilon_1 + \lambda(1|h|2).$$

CHAPTER 13

The Inverse Problem

Usually in quantum mechanics one is given the potential of a system and one wishes to find the energy levels and eigenfunctions if the particle is bound, or the wavefunctions and hence the amount of scattering off this potential if the particle is free. Sometimes however, one knows the wavefunction (or phase shifts) and wishes to find the potential which produces this wavefunction (or these phase shifts). In the literature this is known as the 'inverse' problem.

Consider the one-dimensional eigenvalue equation

$$\left\{-\frac{\hbar^2}{2m}\frac{d^2}{dx^2} + V(x)\right\}\psi(x) = E\psi(x).$$

Solving for V(x) yields

$$\boxed{V(x) = E + \frac{\hbar^2}{2m}\frac{1}{\psi(x)}\frac{d^2\psi(x)}{dx^2} \qquad -\infty < x < \infty} \tag{13.1}$$

EXAMPLE 13.1. Find

$$V(x) \quad \text{for} \quad \psi(x) = \frac{1}{4}\sqrt{\frac{15}{a^5}}\,(a^2-x^2) \qquad |x| < a$$

$$= 0 \qquad |x| > a. \tag{13.2}$$

One evaluates

$$\frac{1}{\psi(x)}\frac{d^2\psi}{dx^2} = \frac{2}{x^2-a^2} \qquad |x| < a.$$

From Equation (13.1) this implies

$$V(x) = E + \frac{\hbar^2}{m(x^2-a^2)} \qquad |x| < a. \tag{13.3}$$

To find V(x) and E individually one must impose an additional condition.

Suppose one is also given the information that $\langle V\rangle = 0$.

This implies

$$E = -\frac{\hbar^2}{m}\frac{15}{16a^5}\int_{-a}^{a}\frac{(a^2-x^2)^2}{x^2-a^2}\,dx = \frac{5\hbar^2}{4ma^2}$$

and

$$V(x) = \frac{5\hbar^2}{4ma^2} + \frac{\hbar^2}{m(x^2-a^2)} = \frac{\hbar^2}{4ma^2}\left\{\frac{5x^2-a^2}{x^2-a^2}\right\} \quad |x| < a$$

$$= \infty \quad |x| > a. \tag{13.4}$$

Figure 13.1 is a plot of the wavefunction in this example (and Figure 13.2 of the resulting potential) and of the very similar function (in this region), $\phi(x) = \sqrt{1/a}\cos \pi x/2a$.

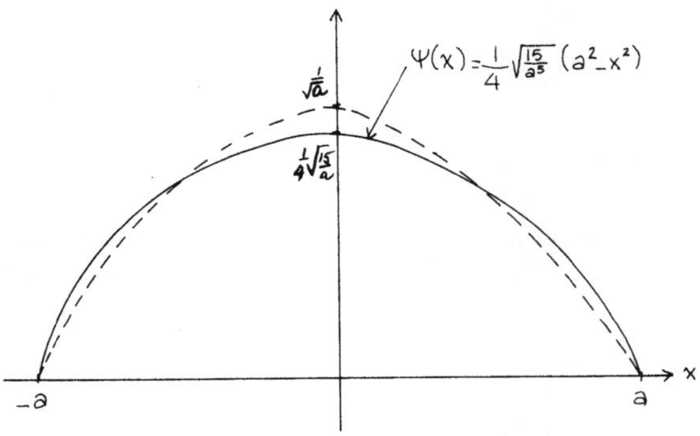

Figure 13.1. Wavefunction in Example 13.1 (full line) and the similar wavefunction $\phi(x) = \sqrt{1/a}\cos \pi x/2a$ (dotted line).

Other conditions, for example that $V(0) = 0$, or $E = 0$ merely shift the energy reference.

Thus if

$$V(0) = 0, \quad E = \frac{\hbar^2}{ma^2} \quad \text{and} \quad V(x) = \frac{\hbar^2}{ma^2}\frac{x^2}{x^2-a^2} \quad |x| < a,$$

while if

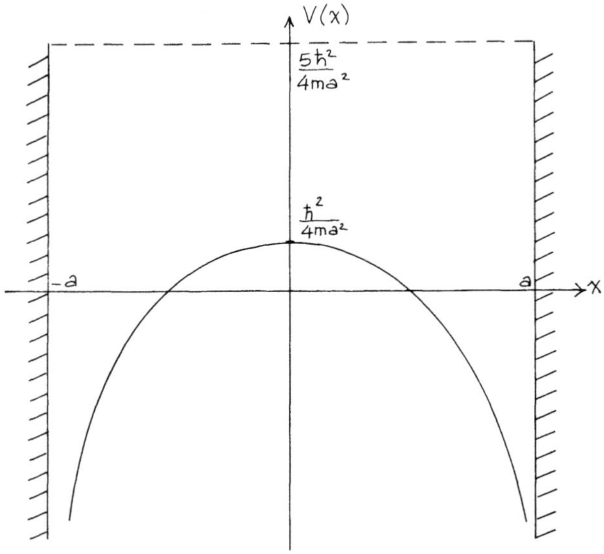

Figure 13.2. Potential in Example 13.1 if <V> = 0.

$$E = 0,\ V(x) = \frac{\hbar^2}{m(x^2-a^2)} \qquad |x| < a.$$

EXAMPLE 13.2. Suppose

$$\psi(x) = \sqrt{2\alpha^3}\ xe^{-\alpha|x|} \tag{13.5}$$

and one wishes to find the potential which results in this wavefunction.

One must evaluate $1/\psi(x)\ d^2\psi(x)/dx^2$. Using the representation

$$\delta(x) = \frac{1}{2}\frac{d^2|x|}{dx^2}$$

for the delta function one obtains

$$\frac{1}{\psi(x)}\frac{d^2\psi(x)}{dx^2} = -2\alpha\delta(x) + \alpha^2 - \frac{2\alpha}{|x|}.$$

Hence

$$V(x) = E + \frac{\hbar^2\alpha^2}{2m} - \frac{\hbar^2\alpha}{m}\left[\frac{1}{|x|} + \delta(x)\right]$$

i.e.

$$V(x) = -\frac{\hbar^2\alpha}{m}\left[\frac{1}{|x|} + \delta(x)\right],\quad E = -\frac{\hbar^2\alpha^2}{2m}. \tag{13.6}$$

The potential here is a sum of a Coulomb potential and a delta function potential (at the origin). As $\psi(x)$ is odd the delta function does not affect this wavefunction's energy which is just that due to the Coulomb potential (see Equation (3.32) with n = 2). This example is illustrated in Figure 13.3.

Three dimensional inverse problems follow along the same lines.

EXAMPLE 13.3. Suppose the wavefunction describing a spinless particle of mass m in a short-range central potential is:

$$\psi(r,\ \theta,\ \phi) = A\,\frac{e^{-\alpha r} - e^{-\beta r}}{r} \tag{13.7}$$

where A, α, β are constants and $\alpha < \beta$.

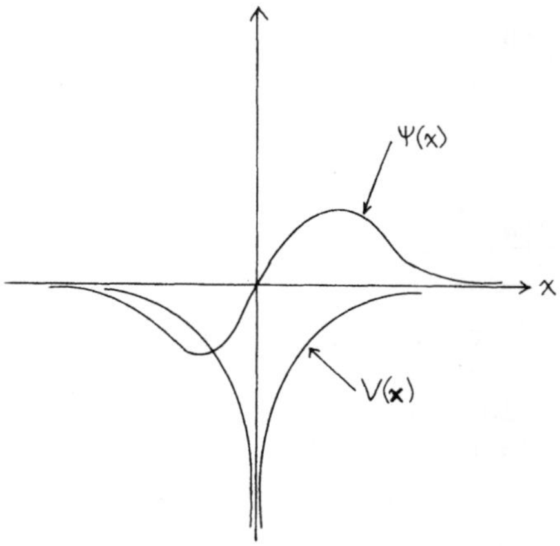

Figure 13.3. Wavefunction and potential of Example 13.2.

From this data find the angular momentum of this particle, the energy of this state and the potential which results in this wavefunction. Firstly

$$L^2(\theta,\ \phi)\psi(r,\ \theta,\ \phi) = 0 = \hbar^2\ell(\ell+1)\psi(r,\ \theta,\ \phi)$$

and

$$L_z(\phi)\psi(r,\ \theta,\ \phi) = 0 = \hbar m\ \psi(r,\ \theta,\ \phi).$$

Thus ℓ, m are zero. Indeed this particular wavefunction has no θ, ϕ i.e. no angular dependence.
Also

$$u_{n0}(r) = rR_{n0}(r) = A(e^{-\alpha r} - e^{-\beta r}).$$

Since the Schrödinger equation for $u_{n0}(r)$ is like that for a one-dimensional problem one can use Equation (13.1) and

$$V(r) = E + \frac{\hbar^2}{2m}\frac{1}{u_{n0}(r)}\frac{d^2u_{n0}(r)}{dr^2} = E + \frac{\hbar^2(\alpha^2e^{-\alpha r} - \beta^2e^{-\beta r})}{2m(e^{-\alpha r} - e^{-\beta r})} .$$

For large r, V(r) goes to zero, i.e.

$$0 = E + \frac{\hbar^2\alpha^2}{2m}$$

(since $\beta > \alpha$), or $E = -\frac{\hbar^2\alpha^2}{2m}$.

Thus generally

$$V(r) = \frac{\hbar^2}{2m}\left[\frac{(\alpha^2-\beta^2)e^{-\beta r}}{e^{-\alpha r} - e^{-\beta r}}\right]. \tag{13.8}$$

For small r,

$$V(r) \approx \frac{\hbar^2}{2m}\frac{(\alpha-\beta)(\alpha+\beta)e^{-\beta r}}{(\beta-\alpha)r} = -\frac{\hbar^2(\alpha+\beta)e^{-\beta r}}{2mr} . \tag{13.9}$$

Equation (13.9) is a 'shielded' Coulomb potential i.e. a potential which looks like a Coulomb potential for small r but which goes to zero much faster than the Coulomb potential for large r.
If $\ell \neq 0$ one easily generalizes Equation (13.1)

$$V(r) = E - \frac{\hbar^2\ell(\ell+1)}{2mr^2} + \frac{\hbar^2}{2m}\frac{1}{u_{n\ell}(r)}\frac{d^2u_{n\ell}(r)}{dr^2}. \quad r \geq 0 \tag{13.10}$$

EXAMPLE 13.4. Suppose

$$\psi(x) = \sqrt{\alpha}\, e^{-\alpha|x|} \tag{13.11}$$

find V(x), E.
Evaluating

$$\frac{1}{\psi(x)}\frac{d^2\psi(x)}{dx^2} = \alpha^2 - 2\alpha\delta(x)$$

(see also Example 13.2).
Thus

$$V(x) = E + \frac{\hbar^2\alpha^2}{2m} - \frac{\hbar^2\alpha}{m}\delta(x).$$

A suitable choice here is

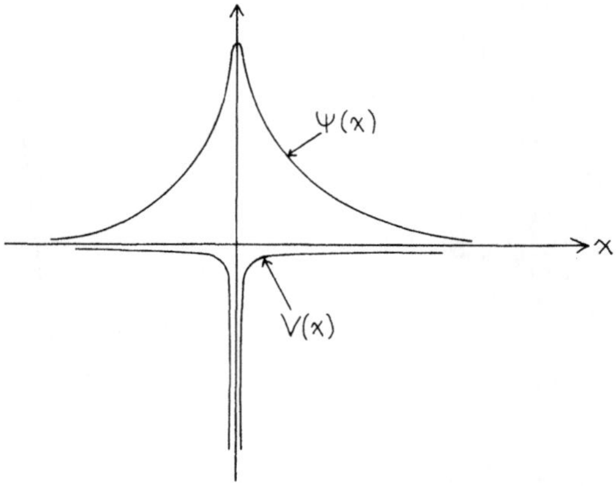

Figure 13.4. Potential and wavefunction in Example 13.4.

$$E = -\frac{\hbar^2}{2m}\alpha^2 \;; \quad V(x) = -\frac{\hbar^2\alpha}{m}\delta(x). \tag{13.12}$$

This problem is illustrated in Figure 13.4.

EXAMPLE 13.5. Suppose

$$\psi(x) = \sqrt{\frac{2}{a\Gamma\left(\frac{1}{4}\right)}}\, e^{-\left(\frac{x}{a}\right)^4}, \tag{13.13}$$

find V(x), E.

Evaluating

$$\frac{1}{\psi(x)}\frac{d^2\psi(x)}{dx^2} = 16\,\frac{x^6}{a^8} - 12\,\frac{x^2}{a^4},$$

$$V(x) = E + \frac{2\hbar^2}{ma^2}\left(\frac{x}{a}\right)^2\left\{4\left(\frac{x}{a}\right)^4 - 3\right\}.$$

The identification

$$V(x) = \frac{2\hbar^2}{ma^2}\left(\frac{x}{a}\right)^2\left\{4\left(\frac{x}{a}\right)^4 - 3\right\}, \quad E = 0, \tag{13.14}$$

is acceptable.
This problem is illustrated in Figure 13.5.

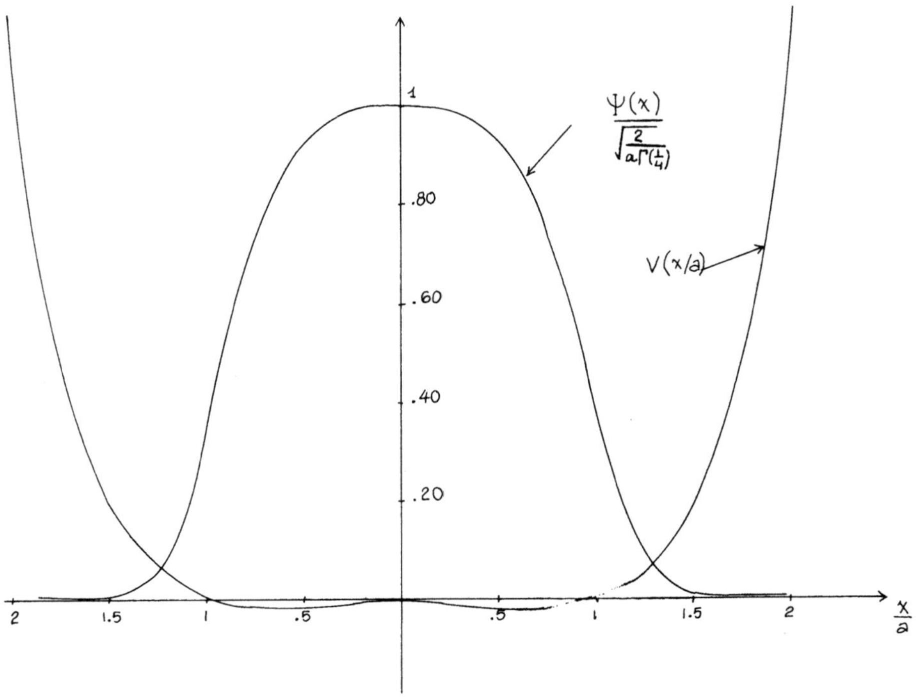

Figure 13.5. Potential and wavefunction in Example 13.5.

CHAPTER 14

The Dalgarno-Lewis Technique

In treating quantum mechanical systems which do not admit to exact solutions for their energy levels and corresponding wavefunctions one has at one's disposal as mentioned in a previous chapter the variational approach (Chapter 10). This however, is restricted to the ground and possibly first excited state (if the Hamiltonian commutes with the parity operator). For other states (and even for the ground state if one requires greater accuracy or additional information against which to juxtapose the variational results) one must fall back on the straightforward application of perturbation theory (Chapter 11) which is generally tedious and at best only approximate. This is because, for other than the first-order results, using perturbation theory one has to evaluate infinite sums in each order (cf. expressions (11.14), (11.15) which one generally approximates by selecting only a few terms which arise to that order.

In this context there does however exist a technique, first pointed out by Dalgarno and Lewis[1]) which in some cases allows one to do away with tedious summations and gives exact answers to a given order.

The basic equation which defines the operator involved, $F_n(x)$, is:

$$[F_n, H_0]\,\phi_n = (h - E_n^{(1)})\phi_n, \tag{14.1}$$

where the Hamiltonian H of a given system is broken up into $H_0 + h$ and the eigenfunctions of H_0 are ϕ_n. Moreover $E_n^{(1)}$ is just the first-order energy term for such a decomposition of the Hamiltonian namely $E_n^{(1)} = (\phi_n h \phi_n)$. The matrix elements obtained from expression (14.1) namely $(\phi_n|[F_n, H_0]|\phi_n)$ are consistent for the diagonal case since

$$(\phi_n|[F_n, H_0]|\phi_n) = 0 = (\phi_n|h - E_n^{(1)}|\phi_n),$$

and for non-diagonal matrix elements

$$(\phi_m|F_n|\phi_n) = \frac{(\phi_m|h|\phi_n)}{\varepsilon_n - \varepsilon_m} \quad (m \neq n). \tag{14.2}$$

Evaluation of the commutator in Equation (14.1) for a one-dimensional Hamiltonian whose potential energy term does not involve the momentum yields:

$$\phi_n \frac{d^2F_n}{dx^2} + 2\,\frac{dF_n}{dx}\,\frac{d\phi_n}{dx} = \frac{2m}{\hbar^2}\,(h - E_n^{(1)})\,\phi_n = \frac{1}{\phi_n}\,\frac{d}{dx}\left(\phi_n^2\,\frac{dF_n}{dx}\right)$$

and with a little manipulation this leads to the closed form expression:

$$F_n(x) = \int^{x} \frac{1}{\phi_n^2(y)} \left\{\int_a^{y} W_n(\xi)\phi_n^2(\xi)\,d\xi\right\} dy. \qquad (14.3)$$

$$W_n(\xi) = \frac{2m}{\hbar^2}\left(h - E_n^{(1)}\right)$$

One notes F_n is clearly state dependent. In expression 14.3, a is a conveniently chosen constant (usually 0 or ∞). Moreover as can be seen from Equation (14.1) or Equation (14.3) F_n is determined only to within an arbitrary constant.

The usefulness of $F_n(x)$ is obvious if one considers for instance the second-order energy term:

$$E_n^{(2)} = \sum_{m\neq n} \frac{V_{nm}V_{mn}}{\varepsilon_n-\varepsilon_m} = \sum_{m\neq n} \frac{(\phi_n|h|\phi_m)(\phi_m|h|\phi_n)}{\varepsilon_n-\varepsilon_m} = \sum_{m\neq n} (\phi_n h\phi_m)(\phi_m F_n\phi_n)$$

and using closure:

$$E_n^{(2)} = (\phi_n|hF_n|\phi_n) - E_n^{(1)}(\phi_n|F_n|\phi_n), \qquad (14.4)$$

a result which involves only two, as opposed to an infinite number of matrix elements.

Similarly the wavefunction to first-order is:

$$\psi_n = N\left(\phi_n + \sum_{m\neq n} \frac{\phi_m V_{mn}}{\varepsilon_n-\varepsilon_m}\right) = N\left\{1+F_n-(\phi_n|F_n|\phi_n)\right\}\phi_n, \qquad (14.5)$$

where N = 1 if one requires only that ψ_n is a cross-normalized function i.e.

$(\psi_n|\phi_n) = 1$, and

$$N = (1 + \Delta F_n)^{-\frac{1}{2}}$$

(where $\Delta F_n = (\phi_n|F_n^2|\phi_n) - (\phi_n|F_n|\phi_n)^2$)
if one requires $(\phi_n|\phi_n) = 1$.

The third-order energy is:

$$E_n^{(3)} = \sum_{m,p\neq n} \frac{V_{nm}V_{mp}V_{pn}}{(\varepsilon_n-\varepsilon_m)(\varepsilon_n-\varepsilon_p)} - E_n^{(1)} \sum_{m\neq n} \frac{V_{nm}V_{mn}}{(\varepsilon_n-\varepsilon_m)^2} =$$

$$= \sum_{m,p\neq n} (\phi_n|F_n|\phi_m)(\phi_m|h|\phi_p)(\phi_p|F_n|\phi_n) -$$

$$- E_n^{(1)} \sum_{m\neq n} (\phi_n|F_n|\phi_m)(\phi_m|F_n|\phi_n) .$$

Using closure here also, one obtains:

$$\begin{aligned} E_n^{(3)} &= (\phi_n|F_nhF_n|\phi_n)-2(\phi_n|F_n|\phi_n)(\phi_n|F_nh|\phi_n)+2E_n^{(1)}(\phi_n|F_n|\phi_n)^2 \\ &\quad -E_n^{(1)}(\phi_n|F_n^2|\phi_n) \qquad (14.6) \\ &= (\phi_n|F_nhF_n|\phi_n) - 2E_n^{(2)}(\phi_n|F_n|\phi_n) - E_n^{(1)}(\phi_n|F_n^2|\phi_n), \end{aligned}$$

a result which needs only two additional integrals for its evaluation.

Some examples which illustrate the efficacy of this technique are given below.

EXAMPLE 14.1. Consider

$$H = \frac{p^2}{2m} + V$$

where

$$V = -\frac{\alpha\hbar c}{x} + \frac{1}{2}m\omega^2x^2 \quad x \geq 0 \ , \quad V = \infty \quad x < 0$$

(and $\alpha = e^2/(4\pi\varepsilon_0\hbar c)$) is the dimensionless fine structure constant $\alpha \approx \frac{1}{137}$ for the Hydrogen atom). Treating $\frac{p^2}{2m} - \frac{\alpha\hbar c}{x}$ as the unperturbed Hamiltonian obtain the Dalgarno-Lewis function F_0. Use this to obtain the energy of the ground state to third order, the wavefunction to first order, and two upper bounds to the energy of this system.

For the partition

$$H_0 = \frac{p^2}{2m} - \frac{\alpha\hbar c}{x} \quad , \quad h = \frac{1}{2} m\omega^2 x^2,$$

$$E_0^{(0)} \equiv \varepsilon_0 = -\frac{1}{2} mc^2\alpha^2, \quad \phi_0 = \sqrt{4\left(\frac{\alpha mc}{\hbar}\right)^3}\; x\, e^{-\frac{\alpha mcx}{\hbar}}$$

and

$$E_0^{(1)} = (\phi_0|\tfrac{1}{2} m\omega^2 x^2|\phi_0) = \frac{3}{2}\left(\frac{\hbar\omega}{mc^2\alpha^2}\right)^2 \alpha^2 mc^2 ,$$

i.e. to first order in perturbation theory

$$E_0^{(0)} + E_0^{(1)} = -\frac{1}{2} mc^2\alpha^2 \left\{1 - 3\left(\frac{\hbar\omega}{mc^2\alpha^2}\right)^2\right\}. \tag{14.7}$$

One notes this series involves the ratio of the energy level parameter ($\hbar\omega$) for the Hamiltonian $H' = p^2/2m + h$ to the energy level parameter for H_0, $(mc^2\alpha^2)$, and as expected the bigger ω the more important the repulsive first-order correction to the energy of this system.

Substituting in Equation (14.3) with the convenient choice $a = \infty$ one obtains:

$$F_0(x) = -\frac{m\omega^2}{6\hbar c\alpha} x^3 - \frac{\omega^2}{2c^2\alpha^2} x^2,$$

which one can easily verify satisfies Equation (14.1).

One readily obtains:

$$(\phi_0|F_0|\phi_0) = -\frac{11}{4}\left(\frac{\hbar\omega}{mc^2\alpha^2}\right)^2$$

and

$$(\phi_0|hF_0|\phi_0) = -\frac{195}{16}\left(\frac{\hbar\omega}{mc^2\alpha^2}\right)^4 mc^2\alpha^2.$$

Hence substituting in Equation (14.4),

$$E_0^{(2)} = -\frac{129}{16} mc^2\alpha^2 \left(\frac{\hbar\omega}{mc^2\alpha^2}\right)^4,$$

a result again directly proportional to (in this case the fourth power of) ω.

To obtain the exact energy to third order one must evaluate the two additional integrals:

$$(\phi_0|F_0hF_0|\phi_0) = \frac{4095}{16}\left(\frac{\hbar\omega}{mc^2\alpha^2}\right)^6 mc^2\alpha^2,$$

and

$$(\phi_0|F_0^2|\phi_0) = \frac{55}{2}\left(\frac{\hbar\omega}{mc^2\alpha^2}\right)^4 .$$

Substituting into Equation (14.6) one obtains

$$E_0^{(3)} = \frac{5451}{32}\left(\frac{\hbar\omega}{mc^2\alpha^2}\right)^6 mc^2\alpha^2 .$$

Thus:

$$E_0^{(0)} + E_0^{(1)} + E_0^{(2)} + E_0^{(3)} = -\frac{1}{2} mc^2\alpha^2 \left\{1 - 3\left\{\frac{\hbar\omega}{mc^2\alpha^2}\right\}^2 + \frac{129}{8}\left\{\frac{\hbar\omega}{mc^2\alpha^2}\right\}^4 - \frac{5451}{16}\left\{\frac{\hbar\omega}{mc^2\alpha^2}\right\}^6 \right\} . \tag{14.8}$$

The wavefunction to first order is:

$$\psi_0 = N\left\{1 + \frac{11}{4}\left(\frac{\hbar\omega}{mc^2\alpha^2}\right)^2 - \frac{\omega^2}{2c^2\alpha^2} x^2 - \frac{m\omega^2}{6\hbar c\alpha} x^3\right\} \phi_0 ,$$

$$N = 1 \text{ or } \left\{1 + \frac{319}{16}\left(\frac{\hbar\omega}{mc^2\alpha^2}\right)^4\right\}^{-\frac{1}{2}} .$$

If one uses as trial wavefunction in Equaton (10.2) the wavefunction ϕ_0, ground state eigenfunction H_0 in the decomposition $H = H_0 + h$ one obtains:

$$E_0 = E_{\substack{\text{ground}\\ \text{exact}}} \leq E_0^{(0)} + E_0^{(1)} . \tag{14.9}$$

If one instead uses for trial wavefunction in Equation (10.2), Equation (14.5) i.e. the wavefunction correct to first order (with $N = (1 + \Delta F_0)^{-\frac{1}{2}}$) one obtains

$$E_0 = E_{\substack{\text{ground}\\ \text{exact}}} \leq E_0^{(0)} + E_0^{(1)} + \frac{E_0^{(2)} + E_0^{(3)}}{1 + \sum\limits_{n\neq 0}^{\infty} \frac{V_{0n}V_{n0}}{(\varepsilon_0 - \varepsilon_n)^2}} = E_0^{(0)} + E_0^{(1)} + \frac{E_0^{(2)} + E_0^{(3)}}{1 + \Delta F_0^2} . \tag{14.10}$$

All the terms needed to evaluate Equation (14.10) are known in this case, yielding

$$E_{\substack{\text{ground}\\ \text{exact}}} \le E_0^{(0)}+E_0^{(1)} - \frac{\left\{\frac{\hbar\omega}{mc^2\alpha^2}\right\}^4\left(\frac{129}{16}-\frac{5451}{32}\left\{\frac{\hbar\omega}{mc^2\alpha^2}\right\}^2\right)mc^2\alpha^2}{1+\frac{319}{16}\left\{\frac{\hbar\omega}{mc^2\alpha^2}\right\}^2} . \qquad (14.11)$$

Both Equations (14.7) and (14.11) are upper bounds to the exact E_{ground}.

EXAMPLE 14.2. Consider the system

$$H = \frac{p^2}{2m} + V$$

$$V = \frac{1}{2} m\omega^2 x^2 + \frac{\alpha}{x^2} \quad x \ge 0 , \quad V = \infty \quad x < 0 .$$

Obtain the energy of this system to second order in perturbation theory using the Dalgarno-Lewis technique.

It is reasonable to partition H into

$$H_0 = \frac{p^2}{2m} + \frac{1}{2} m\omega^2 x^2 \text{ and } h = \frac{\alpha}{x^2} .$$

Then

$$\phi_0(x) = \left(\frac{m\omega}{\hbar}\right)^{3/4} \frac{2x}{\pi^{1/4}} e^{-m\omega x^2/2\hbar} ,$$

$$E_0^{(0)} = \frac{3}{2}\hbar\omega, \; E_0^{(1)} = \frac{2m\omega\alpha}{\hbar} \text{ and } h = \frac{\alpha}{x^2} = \frac{\hbar E_0^{(1)}}{2m\omega x^2} .$$

(If $\alpha = \frac{\hbar^2}{m}$, $E_0^{(0)} + E_0^{(1)} = \frac{7}{2}\hbar\omega$, as opposed to the exact answer in this case (see Example (10.5), namely $\frac{5}{2}\hbar\omega$).

$$F_0 = \int^x \frac{1}{y^2 e^{-m\omega y^2/\hbar}} \int_a^y W(z) z^2 e^{-m\omega z^2/\hbar} dzdy$$

where

$$W(z) = \frac{2m}{\hbar^2}\left(h - E_0^{(1)}\right).$$

With the convenient choice a = 0 and using[2])

$$\int_0^u e^{-u^2} du = e^{-u^2} \sum_{k=0}^{\infty} \frac{2^k u^{2k+1}}{(2k+1)!!} ,$$

$$F_0 = \frac{2E_0^{(1)}}{\hbar\omega} \sum_{k=1}^{\infty} \frac{2^{k-2}\left(\frac{m\omega x^2}{\hbar}\right)^k}{k(2k+1)!!} + \frac{E_0^{(1)}}{2\hbar\omega} \ln \frac{m\omega x^2}{\hbar}$$

$$- \frac{2E_0^{(1)}}{\hbar\omega} \sum_{k=0}^{\infty} \frac{2^{k-1}\left(\frac{m\omega x^2}{\hbar}\right)^{k+1}}{(2k+3)!!(k+1)} = \frac{E_0^{(1)}}{2\hbar\omega} \ln \frac{m\omega x^2}{\hbar}$$

which satisfies Equation (14.1) as one can easily verify. Substituting into Equation (14.4) yields:

$$E_0^{(2)} = \frac{2\left(E_0^{(1)}\right)^2}{\hbar\omega\pi^{\frac{1}{2}}} \left\{ \int_0^{\infty} e^{-u^2} \ln u \, du - 2 \int_0^{\infty} u^2 e^{-u^2} \ln u \, du \right\}$$

$$= \frac{2\left(E_0^{(1)}\right)^2}{\hbar\omega\pi^{\frac{1}{2}}} \left\{ - \frac{\sqrt{\pi}}{4} (\gamma+2\ln 2) + \frac{2\sqrt{\pi}}{8} (\gamma+2\ln 2) - \frac{\sqrt{\pi}}{2} \right\} = \frac{\left(E_0^{(1)}\right)^2}{\hbar\omega} .$$

(where $\gamma = 0.5772157$)

If

$$\alpha = \frac{\hbar^2}{m}, \quad E_0^{(2)} = - 4\hbar\omega , \quad \text{and } E_0^{(0)}+E_0^{(1)}+E_0^{(2)} = \frac{3}{2}\hbar\omega+2\hbar\omega-4\hbar\omega = -\frac{1}{2}\hbar\omega.$$

Substituting into Equation (14.6) one additionally obtains

$$E_0^{(3)} = 2\hbar\omega \left(\frac{E_0^{(1)}}{\hbar\omega}\right)^3.$$

EXAMPLE 14.3. Consider $H = H_0 + h$ where

$$H_0 = \frac{p^2}{2m} + \frac{1}{2} m\omega^2 x^2 , \quad h = \lambda\hbar\omega \sqrt{\frac{m\omega}{\hbar}} \, x \text{ (all x). Evaluate } F_0, E_0^{(2)}$$

etc.

Since

$$\phi_0(x) = \left(\frac{m\omega}{\hbar}\right)^{1/4} \frac{1}{\pi^{1/4}} e^{-m\omega x^2/2\hbar} , \quad E_0^{(0)} = \frac{\hbar\omega}{2} \text{ and } E_0^{(1)} = 0$$

as can easily be shown from parity considerations,

$$F_0(x) = -\lambda \left\{x \sqrt{\frac{m\omega}{\hbar}} \right\},$$

and

$$E_0^{(2)} = -\frac{\lambda^2\hbar\omega}{2}, \quad E_0^{(3)} = 0, \quad \psi = N\left\{1 - \lambda x \sqrt{\frac{m\omega}{\hbar}} \right\}\phi_0$$

one can evaluate the exact energy in this case $= \frac{h\omega}{2} - \frac{h\omega}{2}\lambda^2$ since

$$H = \frac{p^2}{2m} + \frac{\hbar\omega}{2}\left(\sqrt{\frac{m\omega}{\hbar}}x + \lambda\right)^2 - \frac{\hbar\omega}{2}\lambda^2.$$

EXAMPLE 14.4. Given $H = H_0 - h$ where

$$H_0 = \frac{p^2}{2m} + \frac{1}{2}m\omega^2x^2, \quad h = \lambda\hbar\omega\frac{m\omega x^2}{\hbar} \text{ (all x).}$$

Find F_0, $E_0^{(2)}$ etc.

From Example 14.3 ϕ_0 and $E_0^{(0)}$ are known

$$E_0^{(1)} = \lambda\hbar\omega\left(\phi_0\frac{m\omega^2x^2}{2}\phi_0\right)\frac{2}{\hbar\omega} = \lambda\hbar\omega \ \frac{1}{4}\hbar\omega\frac{2}{\hbar\omega} = \frac{\lambda\hbar\omega}{2}.$$

$$F_0 = \frac{\lambda}{2}\left\{\frac{m\omega}{\hbar}\right\}x^2. \quad E_0^{(2)} = -\frac{\lambda^2}{4}\hbar\omega.$$

The exact energy in this case can easily be shown to be

$$\frac{1}{2}\hbar\omega\sqrt{(1+2\lambda)} = \frac{1}{2}\hbar\omega\left\{1 + \lambda - \frac{\lambda^2}{2} + \ldots\right\}$$

EXAMPLE 14.5. Given $H = H_0 + h$ where

$$H_0 = \frac{p^2}{2m} + \frac{1}{2}m\omega^2x^2, \quad h = \lambda\hbar\omega\left(\frac{m\omega}{\hbar}\right)^{3/2}x^3 \text{ (all x),}$$

obtain F_0, and $E_0^{(2)}$.

Using the expressions in Example 14.3 one easily sees $E_0^{(1)} = 0$ from parity considerations, while,

$$F_0^{(1)} = -\lambda\left(\frac{x^3}{3}\left\{\frac{m\omega}{\hbar}\right\}^{3/2} + x\left\{\frac{m\omega}{\hbar}\right\}^{1/2}\right)$$

and $E_0^{(2)} = -\dfrac{11\hbar\omega\lambda^2}{8}$.

EXAMPLE 14.6. Given $H = H_0 + h$ where

$$H_0 = \frac{p^2}{2m} + \frac{1}{2}m\omega^2x^2 \quad \text{and} \quad h = \lambda h\omega\sqrt{\frac{m\omega}{\hbar}}\,x \quad \text{(all x)}$$

obtain $F_1(x)$, $E_1^{(2)}$etc.

$$\phi_1(x) = \left(\frac{m\omega}{\hbar}\right)^{3/2}\frac{\sqrt{2}\,x}{\pi^{1/4}}\,e^{-m\omega x^2/2\hbar}, \quad E_1^{(0)} = \frac{3}{2}\hbar\omega \ .$$

One obtains $E_1^{(1)} = 0$ from parity considerations.

$$F_1(x) = -\lambda\left(\sqrt{\frac{m\omega}{\hbar}}\,x - \frac{1}{x}\sqrt{\frac{\hbar}{m\omega}}\right), \quad E_1^{(2)} = (\phi_1 Fh\phi_1) = -\frac{\lambda^2\hbar\omega}{2}$$

which turns out to be the only non-zero correction to $E_1^{(0)}$ for this system (cf. Example 14.3).

EXAMPLE 14.7. Obtain $F_1(x)$ and $E_1^{(2)}$ if

$$H_0 = \frac{p^2}{2m} + \frac{1}{2}m\omega^2x^2 \quad \text{and} \quad h = \lambda\hbar\omega\left(\frac{m\omega}{\hbar}x^2\right) \quad \text{(all x).}$$

Here

$$\phi_1(x) = \left(\frac{m\omega}{\hbar}\right)^{3/4}\frac{\sqrt{2}\,x}{\pi^{1/4}}\,e^{-m\omega x^2/2\hbar} \text{ and } E_1^{(1)} = \frac{\lambda 3\hbar\omega}{2} \ .$$

In this case

$$F_1(x) = -\frac{\lambda m\omega x^2}{2\hbar}$$

(The same result as for $F_0(x)$ of Example 14.4), an expression which can easily be shown to satisfy Equation (14.1). Using $F_1(x)$ one obtains $E_1^{(2)} = -\frac{3\hbar\omega}{4}\lambda^2$, which agrees with the λ^2 term in the expansion of the exact energy $\frac{3}{2}\hbar\omega\,(1 + 2\lambda)^{1/2}$ in powers of λ. (The $F_n(x)$'s for this system if $n > 1$ are more complicated than $F_1(x) = F_0(x)$).

A useful property of the F function is that it is additive. Thus if for a given H_0, h_1, and h_2 individually result in first-order energies

$E_n^{1(1)}$, $E_n^{2(1)}$ and lead to F_n^1 and F_n^2 respectively, for the case $h = h_1 + h_2$, the defining equation (1) is satisfied by $F_n^1 + F_n^2$ since $E_n^{(1)} = E_n^{1(1)} + E_n^{2(1)}$. One can thus use results from simpler problems in more complex situations.

EXAMPLE 14.8. Given $H_0 = \frac{p^2}{2m} + \frac{1}{2} m\omega^2 x^2$ and

$$h = \lambda_1 \hbar\omega \sqrt{\frac{m\omega}{\hbar}}\, x + \lambda_2 \hbar\omega \left(\sqrt{\frac{m\omega}{\hbar}}\, x\right)^2 \quad \text{(all x)}.$$

Find F_0, F_1 etc.

Combining the results of Examples 14.3 and 14.4.

$$F_0 = -\lambda_1 \left\{\sqrt{\frac{m\omega}{\hbar}}\, x\right\} - \frac{\lambda_2}{2}\left\{\sqrt{\frac{m\omega}{\hbar}}\, x\right\}^2$$

and from Examples 14.6 and 14.7.

$$F_1 = -\lambda_1 \left\{\sqrt{\frac{m\omega}{\hbar}}\, x - \frac{1}{x}\sqrt{\frac{\hbar}{m\omega}}\right\} - \frac{\lambda_2}{2}\left\{\sqrt{\frac{m\omega}{\hbar}}\, x\right\}^2$$

$$E_0^{(1)} = \lambda_2 \hbar\omega/2 \qquad E_1^{(1)} = 3\lambda_2 \hbar\omega/2$$

$$E_0^{(2)} = -\hbar\omega\left(\frac{\lambda_1^2}{2} + \frac{\lambda_2^2}{4}\right) \qquad E_1^{(2)} = -\hbar\omega\left(\frac{\lambda_1^2}{2} + \frac{3\lambda_2^2}{4}\right)$$

i.e.

$$E_0 = \frac{\hbar\omega}{2}\left\{1 + \lambda_2 - \left(\lambda_1^2 + \frac{\lambda_2^2}{2}\right) + \ldots\right\}$$

$$E_{\text{1st existed}} = \frac{3\hbar\omega}{2}\left\{1 + \lambda_2 - \left(\frac{\lambda_1^2}{3} + \frac{\lambda_2^2}{2}\right) + \ldots\right\}$$

in agreement with the exact energy results:

$E_{\text{exact ground}}$

$$= \frac{1}{2}\hbar\omega\sqrt{1 + 2\lambda_2} - \frac{\hbar\omega\lambda_1^2}{2(1+2\lambda_2)}$$

$E_{\text{exact first excited state}}$

$$= \frac{3}{2}\hbar\omega\sqrt{1+2\lambda_2} - \frac{\hbar\omega\lambda_1^2}{2(1+2\lambda_2)},$$

expanded to order λ^2.

EXAMPLE 14.9. Consider the Hamiltonian

$$H = \frac{p^2}{2m} - \frac{\alpha\hbar c}{x} + \frac{\hbar^2\beta}{2mx^2} \qquad x \geq 0 .$$

Treating $h = \hbar^2\beta/2mx^2$ as a perturbation, study the energy of the ground state of this system. The unperturbed ground-state wavefunction is

$$\phi_0 = \sqrt{4\left(\frac{\alpha mc}{\hbar}\right)^3} \; xe^{-\frac{\alpha mcx}{\hbar}}$$

and the unperturbed ground-state energy is

$$E_0^{(0)} = -\frac{1}{2} mc^2\alpha^2 .$$

One immediately evaluates the first-order energy contribution

$$E_0^{(1)} = \left(\phi_0 \left| \frac{\hbar^2\beta}{2mx^2} \right| \phi_0\right) = \beta\alpha^2 mc^2 .$$

The exact energy if $\beta = 2$ is $-\frac{1}{8}\alpha^2 mc^2$, which ($E_0^{(0)} + E_0^{(1)}$ being an upper bound) is less than $(-\frac{1}{2} + 2)\alpha^2 mc^2 = \frac{3}{2}\alpha^2 mc^2$.
Using Equation (14.3) one obtains F_0 in this case: (with $a = \infty$)

$$F_0 = \frac{mc\alpha\beta}{\hbar} x + \beta \ln \frac{2\alpha mcx}{\hbar} .$$

Hence evaluating Equation (14.4) in this case yields:

$$E_0^{(2)} = -\frac{5}{2}\beta^2\alpha^2 mc^2 ,$$

where one uses the fact

$$\int_0^\infty e^{-u} \ln u \, du = -\gamma \text{ etc.}$$

Using Equation (14.6) one can also evaluate $E_0^{(3)}$,

$$E_0^{(3)} = 7\beta^3 mc^2\alpha^2 .$$

REFERENCES

1. A. Dalgarno and J.T. Lewis, Proc. Roy. Soc. A233(1955) 70, C. Schwartz, Ann. Phys. 6, (1959), 156.
2. I.S. Gradshteyn, Academic Press (1980) p. 306.

Index

GPSR Compliance

The European Union's (EU) General Product Safety Regulation (GPSR) is a set of rules that requires consumer products to be safe and our obligations to ensure this.

If you have any concerns about our products, you can contact us on ProductSafety@springernature.com

In case Publisher is established outside the EU, the EU authorized representative is:

Springer Nature Customer Service Center GmbH
Europaplatz 3
69115 Heidelberg, Germany

Batch number: 10607330

Printed by Printforce, the Netherlands